ACS SYMPOSIUM SERIES **554**

Emerging Technologies in Hazardous Waste Management IV

D. William Tedder, EDITOR
Georgia Institute of Technology

Frederick G. Pohland, EDITOR
University of Pittsburgh

Developed from a symposium sponsored
by the Division of Industrial and Engineering Chemistry, Inc.,
of the American Chemical Society
at the Industrial and Engineering Chemistry Special Symposium,
Atlanta, Georgia,
September 21–23, 1992

American Chemical Society, Washington, DC 1994

Library of Congress Cataloging-in-Publication Data

Emerging technologies in hazardous waste management IV / D. William Tedder, editor, Frederick G. Pohland, editor.

p.   cm.—(ACS symposium series, ISSN 0097–6156; 554)

"Developed from a symposium sponsored by the Division of Industrial and Engineering Chemistry, Inc., of the American Chemical Society at the Industrial and Engineering Chemistry Special Symposium, Atlanta, Georgia, September 21–23, 1992."

Includes bibliographical references and indexes.

ISBN 0–8412–2857–4

1. Hazardous wastes—Management—Congresses.   2. Soil remediation—Congresses.

I. Tedder, D. W. (Daniel William), 1946–   . II. Pohland, F. G. (Frederick George), 1931–   . III. American Chemical Society. Division of Industrial and Engineering Chemistry. IV. Series.

TD1020.E443   1994
628.4′2—dc20   93–50681
                                                                      CIP

Foreword

THE ACS SYMPOSIUM SERIES was first published in 1974 to provide a mechanism for publishing symposia quickly in book form. The purpose of this series is to publish comprehensive books developed from symposia, which are usually "snapshots in time" of the current research being done on a topic, plus some review material on the topic. For this reason, it is necessary that the papers be published as quickly as possible.

Before a symposium-based book is put under contract, the proposed table of contents is reviewed for appropriateness to the topic and for comprehensiveness of the collection. Some papers are excluded at this point, and others are added to round out the scope of the volume. In addition, a draft of each paper is peer-reviewed prior to final acceptance or rejection. This anonymous review process is supervised by the organizer(s) of the symposium, who become the editor(s) of the book. The authors then revise their papers according to the recommendations of both the reviewers and the editors, prepare camera-ready copy, and submit the final papers to the editors, who check that all necessary revisions have been made.

As a rule, only original research papers and original review papers are included in the volumes. Verbatim reproductions of previously published papers are not accepted.

M. Joan Comstock
Series Editor

Contents

Preface

HEALTH AND ENVIRONMENTAL PROBLEMS associated with hazardous wastes and their management continue to challenge society in ever-increasing dimensions. Public concern over the discovery of contaminated soils, sediments, and groundwater aquifers due to past hazardous waste disposal practices has heightened, and remedial activity continues at a rapid pace. These activities, however, have not yielded completely satisfactory solutions because of the difficulty of applying conventional treatment technologies to waste sources in uncertain and complex environmental media.

The symposium on which this volume is based attracted a diverse audience specializing in technologies for hazardous waste management. From the approximately 242 presentations, 100 final manuscripts were submitted for review. The 18 chapters chosen for this volume were selected on the basis of peer review, scientific and technical merit, innovation, and the editors' perceptions of applicability and significance on both a short- and long-term basis.

This volume is a continuation of the *Emerging Technologies in Hazardous Waste Management* theme initiated with ACS Symposium Series No. 422 in 1990, No. 468 in 1991, and No. 518 in 1993, all of which focus on developments related to advances in hazardous waste management at the time. The content of this volume extends into complementary areas with new information and applications not previously emphasized. Highlighted are the changing diverse approaches to some of the most challenging current problems and their solutions. This volume offers an overall introduction to the field of hazardous waste management within the various environmental phases and affords an opportunity to read works by experts possessing broad disciplinary credentials.

Acknowledgments

Several sponsors provided support for the symposium on which this volume is based. We specifically acknowledge with grateful appreciation Merck & Co., Inc.; the Ford Motor Company; and the National Registry of Environmental Professionals, represented by Valcar A. Bowman. In addition, we thank Wallace W. Schulz for his assistance in the peer review process.

D. WILLIAM TEDDER
Georgia Institute of Technology
Atlanta, GA 30332–0100

FREDERICK G. POHLAND
University of Pittsburgh
Pittsburgh, PA 15261–2294

December 3, 1993

Chapter 1

Emerging Technologies for Hazardous Waste Management

An Overview

D. William Tedder[1] and Frederick G. Pohland[2]

[1]School of Chemical Engineering, Georgia Institute
of Technology, Atlanta, GA 30332-0100
[2]Department of Civil and Environmental Engineering,
University of Pittsburgh, Pittsburgh, PA 15261-2294

Hazardous waste management and the remediation of contaminated environments continue to command priority attention and the mobilization of the collective capabilities of scientists, engineers and related disciplines in the private and public sectors of society. This emphasis, and the recognition of a need to efficiently and economically resolve hazardous waste management problems, has led to an emergence of an array of innovative technologies with applications in a variety of environmental settings. Indeed, the focus on hazardous waste management and remediation technology has been the subject of numerous national and international workshops, symposia, conferences and associated activities, as exemplified by the Wastech® initiative of the American Academy of Environmental Engineers(1).

Our current contribution to the advancement of knowledge and its application to the solution of hazardous waste management challenges focuses on selected technologies currently under development, and showing particular promise in terms of new approaches as well as advancing the application of fundamentals of science and technology. As such, it is an extension of and an embellishment to the continuum of topics presented in previous volumes of the series (2-4), with an attempt to highlight those that have taken on particular contemporary significance. Therefore, perusal of this overview will serve to demonstrate the importance of soils and sediments as crucial remediation horizons, the inextricable linkage between waste minimization and management technologies, and the renewed emphasis on the pervasive challenges of radioactive wastes.

Remediation Technologies for Soils and Sediments

The discovery and environmental impacts of contaminated soils and sediments have catalyzed an aggressive search for innovative technologies for both access and remediation, employing physical, chemical and/or biological techniques. In Chapter 2, Chesnut recognizes the consequences of the usual heterogeneity encountered in porous geologic media, including effects on porosity, permeability, mineralogy, and contaminant concentrations, and presents a model for quantifying effects of heterogeneity on vapor extraction, where a single parameter, σ, is used. The resulting model is sufficiently simple to apply in Monte Carlo and similar sampling procedures for estimating the

0097–6156/94/0554–0001$08.00/0

uncertainty in cost or duration of a remediation project. The model, originally developed to determine the effect of heterogeneity on waterflooding for secondary oil recovery, has also been extended elsewhere (5) to estimate travel time distribution for radionuclides dissolved in water moving through fractured porous rock.

Recognizing that soils and groundwater may be contaminated with heavy metals, requiring extensive excavation for removal and management, Lingren, et al. (Chapter 3) developed and tested an electrokinetic process for *in situ* remediation of anionic contaminants in terms of electromigration rate through unsaturated media at varying moisture contents. Decreases in rate at low and high moisture contents were explained by increases in pore tortuosity and decreases in pore water current density, respectively. Experiments with food dye and chromate ions provided the basis for developing a simple experimental model capable of predicting dependence on soil moisture content and pore water chemistry.

Knowledge of biodegration kinetics can also facilitate decisions on *in situ* remediation of soils, sediments and aquifers. Tabak, et al. (Chapter 4) investigated adsorption/desorption equilibria and kinetics in soil slurry reactors for phenol and alkyl phenols. A mathematical model incorporating effects of adsorption/desorption and biodegradation in the liquid and solid soil phases, as well as protocols for measuring biomass adsorption in soil and bioconversion by radiolabelled carbon dioxide evolution, were presented and advocated for use in determining the extent of bioremediation at contaminated sites. As such, this contribution complements the overview presented by Eckenfelder and Norris (6) in the previous volume of this symposium series.

Tsang, et al. (Chapter 5) also focused on the need for re-mediation of soils and sediments contaminated by activities such as mining, electroplating and other manufacturing and industrial processes. Sterile and nonsterile soil, experimentally contaminated with bismuth, cadmium, lead, thorium and uranium, containing cysteine, glycine, or thioglycolate, and inoculated with pure cultures of soil bacterial isolates, indicated the promotion of releases of all contaminants. It was concluded that viable and active microorganisms influence the ability of soil to retain or release metals, and that cysteine is effective in this process, operating as a reducing agent as well as a metal complexing agent and nutrient. Enhancement of microbially-mediated removal was considered due to changes in pH and/or Eh near the soil colloid, alteration of the valence state of the metals, and decomposition of the organic matter in the soil. These findings again emphasize the necessity of providing favorable growth conditions supportive of biological activity on a continuum.

Waste Minimization and Management Technologies

The recognition of waste minimization as an essential element in the hierarchial approach to hazardous waste management continues to foster developments in responsive methodologies. Edgar and Huang (Chapter 6) advocated a systematic module-based synthesis approach to the design of environmentally clean processes with minimal waste generation. The approach adds the dimension of structural controllability to the conventional capital and operating cost functions, and elaborates waste minimization strategies as constraints. In the absence of sufficient data on waste generation at the process design step, artificial intelligence techniques are used to represent waste minimization strategies and evaluate controllability. The efficacy of the proposed waste minimization approach is illustrated using a phenol-containing waste in an oil refinery as an example. Therefore, the generic features of the approach would enhance its applicability to other industrial settings.

Recognizing the increasing difficulty of meeting more stringent effluent quality requirements by conventional treatment technologies, Gaarder, et al. (Chapter 7) selected a crystallization technology for the concentration of mechanical pulp mill effluents, and the subsequent recovery and reuse of water. After screening other separation technologies, clathrate hydrate concentration, a variant of freeze concentration, was selected using carbon dioxide and propane to control the temperature and pressure conditions at which hydrates form. The results of the studies were considered essential to the development of large-scale crystallization units applied to pulp mill effluents where impurities did not adversely affect pressure-temperature equilibria.

The control of mercury in aquatic environments and sediments continues to receive attention, and methods for its removal and recovery have been a subject of considerable scientific inquiry. Larson and Wiencek (Chapter 8) review the various management techniques and advocate the utilization of a microemulsion containing a cation exchanger to extract mercury from water. In order to successfully model the extraction process, both the equilibrium and kinetics of the metal:liquid ion exchange were determined, and a diffusion/reaction model for extraction of mercury with oleic acid in a batch STR (stirred reactor) is presented.

Many hazardous waste constituents are volatile as well as biodegradable, and various methods to capture and convert these fractions have emerged. Apel, et al. (Chapter 9) have focused on the use of laboratory and full-scale biofilters for remediation of gasoline vapors using a proprietary bed medium. Removal occurred over temperature ranges of 22 to 40° C, and removal rate was a function of gasoline concentration and microbial viability. Removal of benzene in field tests was approximately 10 - 15%, with total BTEX (benzene, toluene, ethylbenzene, xylene) removal of 50 - 55%. Optimum contact opportunity between the gas stream and microbial consortia was considered essential and an important design variable.

The recovery and processing of petroleum produces organic and oily wastes requiring appropriate management to comply with regulations. Majid, et al. (Chapter 10) focused on the feasibility of introducing finely divided sulphur dioxide capture agents into high sulphur fuels to demonstrate sulphur capture during combustion process. By using liquid phase agglomeration and comparison of the results of static muffle furnace combustion tests at 850° C with bench-scale fluidized bed results, cogglomeration of petroleum cokes and lime resulted in sulphur capture of over 80 to 30%, depending on the source of carbon and respective test conditions. A Ca:S molar ratio in the range of 1 to 2 was capable of reducing SO_2 emissions sufficiently to meet USA and Canadian standards.

Organic aqueous contaminants may exist as nonionic species and may be treated by sorption and subsequent partitioning for removal and/or recovery. Park and Jaffe (Chapter 11) recognized the ability of anionic surfactant monomers to adsorb onto mineral oxides (organo-oxide) and act as a sorbent for nonionic organic pollutants, with the advantage of *in situ* regeneration. Using batch and column experiments with aluminum oxide, carbon tetrachloride and an anionic surfactant, sorption was shown to be highly pH-dependent, with partitioning linearly dependent on concentration and proportional to the adsorbed mass of surfactant.

High energy election-beam irradiation is also an emerging technology for removing hazardous organic contaminants from water. To explore the effectiveness of this technique, Rosocha, et al. (Chapter 12) formulated a simple kinetic model for removing TCE and CCl_4 from water, examining the production, recombination and reaction of free radicals. Simulations indicated that low pulse intensities were more efficient in producing radicals, however, a train of short, high-dose, repetitive pulses could approach the removal efficiency of a continuous dose profile. Therefore, repetitively pulsed accelerators were considered preferable for future applications because of specific machine advantages.

The ubiquitous distribution of pesticides in the USA and throughout the world has focused attention on their environmental and health effects. Pentachlorophenol (PCP) is the second most widely-used pesticide in the USA and, because of its biocidal properties, it has many industrial applications. Therefore, focus on effective methods of conversion and removal from environmental media remains high. Carberry and Lee (Chapter 13) used Fenton's Reagent to enhance degradation potential for PCP through partial chemical oxidation followed by biological oxidation with selected microbial consortia. Chemical oxidation at a level of only 3% tripled the rate of biological conversion and reduced the toxicity index to a level four times less. It was concluded that the mechanism of enhancement was due to a reduction in PCP toxicity and co-metabolism of the PCP and partial oxidation products.

Pesticides can also be oxidized with ozone, and Hapeman (Chapter 14) studied the aqueous ozonation of s-triazine pesticides and the dependence of conversion product formation structure and abundance on the duration of ozonation. Using chemical methods, HPLC, mass and NMR spectroscopy, a proposed degradation pathway for s-triazines was described, together with a reaction product profile of atrazine. The technique was considered of value in developing methods for triazine residue removal from ground and surface waters.

Radioactive and Mixed Waste Management

The storage and long-term management of nuclear wastes remains a difficult challenge, and increased attention is being given to the assessment and maintenance of facilities at existing sites. Reynolds and Babad (Chapter 15) have described the Hanford Site complex and its facilities for storage of radioactive wastes. Of the 177 existing tanks for storage of radioactive wastes, 53 have been identified as having potential safety issues. The monitoring for generation and release of hydrogen from one of the tanks and minimization of the risk of a hydrogen burn are presented, together with steps for mitigation of hydrogen release.

The storage challenges at the Hanford complex are more explicitly elucidated by mechanistic studies on the thermal chemistry of simulated aqueous nuclear waste mixtures by Ashby, et al. (Chapter 16). This incisive and detailed study of factors affecting the reaction by which H_2 gas is formed from the degradation product, formaldehyde, and OH^- provides a mechanism consistent with results obtained from radiolabelling experiments. Moreover, the amount of hydrogen produced was found to be substantially larger under oxygen than under an argon atmosphere, which has new and important ramifications with regard to mitigation procedures that would introduce O_2 into the waste mixture and thereby intensify the rate and concern associated with H_2 formation.

Vitrification is an innovative technology applicable for the immobilization of high-level radioactive isotopes in borosilicate glass. Bannochie, et al. (Chapter 17) present results of bench-scale acid hydrolysis tests for defining factors affecting removal of aromatic carbon from aqueous slurries of Cs-137 and other tetraphenylborates prior to vitrification at the Savannah River Site. Since hydrolysis is performed with formic acid and Cu(II) as a catalyst, the eventual rate of conversion to the primary products of benzene and boric acid is influenced by the need to balance the Cu(II) and total acid ratio, and the presence or absence of oxygen, nitrite, complexing agents and other potential reaction catalysts or inhibitors.

Solidification and stabilization of hazardous and radioactive wastes are being advanced as viable management techniques. Darnell (Chapter 18) presents the potential of sulfur polymer cement (SPC) or sulfur polymer cement concrete (SPCC) as final waste forms for such applications. Based on a summary of a series of tests in the USA and abroad, guidance is provided for applications of the technology, including a discussion of advantages and disadvantages in terms of waste type and exposure condition.

Summary

The control and remediation of hazardous and radioactive wastes, and their environmental and health consequences, continue to challenge scientific and engineering ingenuity as environmental regulations become more stringent and industrial productivity continues to accelerate commensurate with the needs and aspirations of a growing consumer population. Therefore, it is appropriate that new and innovative management techniques are sought and developed for application, including emphasis on waste minimization and pollution prevention. Indeed, the transformation of an end-of-the-pipe approach to one that prevents pollution and promotes innovation in design and processing, as well as resource conservation and recycle, has shifted much attention to problems involving environmental and health impairment due to past waste management practices. Accordingly, the new emerging technologies in hazardous waste management have become consequences of the challenges of this focus on situations where waste source and environmental setting are more obscure and less explicitly defined. Therefore, it is appropriate to embrace such facets of the challenge, whether related to soils and sediments as a target matrix, waste minimization and *ex situ* management, or the special characteristics of a segregated waste requiring isolation and/or long-term storage.

Each of these broader generic topics is addressed with specific selected examples in this volume of the series, and while definite progress is being made in all cases, additional improvements are necessary, whether in the basic science or technology, or its application in practice. As developments proceed both reactively as a consequence of regulatory inertia, but also proactively and in an anticipatory fashion, a change in protocol will be required, including the development and nurturing of financial and personnel resources capable of managing and sustaining the technology. Therefore, it is anticipated that these issues will collectively contribute, not only to the substance of continuing symposia in print, but to the overall advancement and transfer of the knowledge needed to sustain progress in hazardous waste management in the future.

Literature Cited

1.   Anderson, W. C., Wastech® - Assessing Innovative Waste Treatment Technology. *Waste Management*. 13, 205-206, 1993.
2.   Tedder, D. W. and Pohland, F. G., Eds. *Emerging Technologies in Hazardous Waste Management, ACS Symposium Series 422*, American Chemical Society, Washington, DC, 1990.
3.   Tedder, D. W. and Pohland, F. G., Eds. *Emerging Technologies in Hazardous Waste Management II, ACS Symposium Series 468*, American Chemical Society, Washington, DC, 1991.
4.   Tedder, D. W. and Pohland, F. G., Eds. *Emerging Technologies in Hazardous Waste Management III, ACS Symposium Series 518*, American Chemical Society, Washington, DC, 1993.
5.   Chesnut, D. A., Characterizing the Altered Zone at Yucca Mountain; The Beginning of a Testing Strategy. *Proceedings Third International High-Level Radioactive Waste Management Conference*, 1026-1039, American Nuclear Society, Inc., Las Vegas, NV, 1993.

6.   Eckenfelder, W. W., Jr. and Norris, R. D., Applicability of Biological Processes for Treatment of Soils. In Tedder, D. W. and Pohland, F. G., Eds., *Emerging Technologies in Hazardous Waste Management III, Chapter 8, pages 138-158, ACS Symposium Series 518*, American Chemical Society, Washington, DC, 1993.

RECEIVED December 3, 1993

REMEDIAL TECHNOLOGIES FOR SOILS AND SEDIMENTS

Chapter 2

Heterogeneity and Vapor Extraction Performance

Dwayne A. Chesnut

Earth Sciences Division, Physical Sciences Department, University of California, Lawrence Livermore National Laboratory, Livermore, CA 94550

Soil venting or vapor extraction, the "pump and treat" method of aquifer remediation, and petroleum production by primary or secondary recovery all involve the flow of fluids of varying composition in porous geologic media. Properties of these media which govern fluid flow and contaminant transport, such as porosity, permeability, mineralogy, saturations, and solute concentration, vary from point to point, often by orders of magnitude. As this spatial heterogeneity increases, the performance of the remediation or petroleum production process deteriorates markedly, requiring the extraction of larger and larger total quantities of fluid to recover the desired amount of "product" (contaminant, oil, etc.). A simple model is presented for quantifying the effects of heterogeneity on vapor extraction. It was originally developed to determine the effect of heterogeneity on waterflooding for secondary oil recovery and subsequently extended to estimate the travel time distribution for radionuclides dissolved in water moving through fractured porous rock. Heterogeneity is quantified by a single parameter, σ, which determines the shape of a breakthrough curve for the transport of a tracer or other distinguishable fluid front as it moves through the medium. The resulting extraction model is simple enough to use in Monte Carlo and similar sampling procedures to estimate the uncertainty in cost or duration of the remediation project.

Although differing in important details, modeling groundwater or vadose zone remediation is conceptually similar to modeling enhanced oil recovery, in that both typically involve multi-phase, multi-component fluid systems, transient flow and transport conditions, and complex chemical, physical, and even biological processes governing the transfer of mass and energy among elements of the subsurface system (i.e., the aquifer or petroleum reservoir and its contained fluids) and between the

0097–6156/94/0554–0008$08.90/0

system and its surroundings. Both types of processes involve the displacement of one fluid by another, and fluid displacement efficiency is profoundly affected by the spatial heterogeneity of the geologic media in which these displacements are conducted. Process efficiency, in turn, depends strongly upon fluid displacement efficiency.

Geologic materials are generally much more heterogeneous than the porous media familiar to chemical engineers, such as those used in packed columns. One of the daunting problems in modeling fluid displacement processes in natural materials is the adequate characterization of their heterogeneity. The degree of success in predicting the flow and transport behavior of real systems depends upon how well the actual spatial variability of physical properties is approximated in the model, as well as upon how completely the physical and chemical processes important to process performance are represented.

The degree of variability and its spatial scale depend upon the specific property being estimated. For example, the intrinsic permeability of geologic media can easily vary over five or six orders of magnitude even within distances of a few tens of feet or less, but the fractional porosity is physically constrained to lie between zero and unity. In turn, these different properties have varying impact on flow and transport through the system.

Petroleum Production

The petroleum industry has produced an enormous literature of laboratory investigations, field data, and model studies concerned with the injection of water, carbon dioxide, air, hydrocarbon gases, steam, polymer solutions, and surfactants into petroleum reservoirs, and the resulting production of oil, other native fluids, and injected substances. Much of this work has been focused on the prediction of oil recovery as a function of time, and frequent reality checks between models and experiments have been impartially imposed by economics. Hence, the analysis of petroleum production processes, and appropriate modification of the methods used, may accelerate the development of cost-effective characterization and modeling approaches for remediation projects.

In this paper, we summarize in some detail the development of a simple "heterogeneity parameter" model (*1*) for predicting (or extrapolating) the performance of waterfloods in depleted petroleum reservoirs. The approach is extended to develop a model for contaminant removal, and then applied to the analysis of a pilot project for the extraction of TCE from the vadose zone at the Savannah River Site. A similar analysis (*2*) was used recently to define more precisely the concept of "groundwater travel time" introduced by the Nuclear Regulatory Commission in connection with determining the suitability of a potential repository for high-level nuclear waste.

Primary Recovery.   In petroleum production, the primary recovery stage comprises the removal of oil, gas, and water from one or more wells (or producers) by reducing the bottom-hole pressure at the well-bore to a value below the pressure in the surrounding porous and permeable medium (the petroleum reservoir). In some reservoirs, the initial pressure, producing depth, and fluid density are such that the produced fluids will flow to the surface, but generally some form of artificial lift (e.g., a pump) is required before the end of primary production.

Many reservoirs have essentially closed outer boundaries, with little or no mass entering the system to replace fluids as they are produced. Hence, the reservoir pressure gradually declines in proportion to the cumulative fluid withdrawn (i.e., with increasing *reservoir voidage*), and they are known as *depletion reservoirs*. Economic primary production terminates when the pressure has declined to the point that oil and gas can no longer flow into the producers at rates sufficiently high to provide net revenues greater than the operating costs of the wells.

After a depletion reservoir has reached its economic limit on primary production, a significant percentage of the oil originally in place remains in the rock, along with gas, which exsolves as the reservoir pressure is reduced, and along with much of the water originally present. A portion of the remaining oil, known as the *waterflood movable oil*, can be recovered by water displacement; the remainder is the *waterflood residual oil*, trapped by capillary forces within the rock pores.

In the laboratory, when water is injected continuously into a rock core containing oil at a saturation higher than the waterflood residual value, some of the oil will be displaced to the outflow end of the core. The residual oil saturation value depends upon the initial oil saturation, the flow velocity, and various intrinsic properties of the rock/fluid system. The difference between the initial oil saturation, S_{oi}, and the waterflood residual oil saturation, S_{orw}, is defined as the movable oil saturation. The residual oil after waterflooding represents the target for tertiary recovery processes, such as surfactant flooding.

Waterflooding. In field applications of waterflooding, water is injected into the reservoir, either by converting some of the original producers to injectors, or by drilling new injectors, while continuing to withdraw fluids (usually by pumping) from the remaining producers. The injection of water not only restores the reservoir pressure, it also immiscibly displaces (or *sweeps*) oil from the injectors to the producers. Usually, the array of injectors and producers is designed to form a repeated pattern of basic symmetry elements, with the "five-spot" perhaps being the most common. Except near the edges of the reservoir, each injector in a five-spot pattern flood is surrounded by four producers, located at the corners of a square centered on the injector. A schematic illustration of a single symmetry element in a five-spot pattern is shown in Figure 1. A typical distance between producing wells in the U.S. is about 400 m.

The efficiency of a waterflood is most conveniently expressed as the fractional recovery of the total movable oil remaining in the reservoir after primary depletion. This allows the direct comparison of actual flood performance with an ideal system in which injected water volumetrically displaces, first, the gas left behind at the end of primary production, and then, the movable oil. No oil is produced until enough water has been injected to displace all the gas (the *fill-up* volume). Subsequently, oil alone is produced until water breaks through to the producing well, at which point the cumulative water injected is called the *flood-out* volume.

Ideal Waterflood Performance. This ideal, "piston-like" displacement process can be described quantitatively through application of the following assumptions:

1. Before injection begins, the system is completely depleted by primary production, and there is no flow from the outflow end.
2. Fractional porosity (ϕ), along with initial oil, gas, and water saturations (expressed as fractions of the pore volume) S_{oi}, S_{gi}, and S_{wi}, respectively, are uniform, as is the intrinsic permeability, k.
3. Beginning at time $t = 0$, water is injected at constant rate i_w into one end of the system (the *inflow boundary*), successively displacing gas, oil, and water out the other end (the *outflow boundary*).
4. Flow between the inflow and outflow boundaries is perfectly linear.
5. All displacements of one fluid by another are absolutely stable (*i.e., piston-like*), unaffected by gravity, capillarity, viscous fingering, dispersion, or any other process causing mixing or simultaneous flow of displacing and displaced fluids.
6. Injected water builds up an *oil bank* ahead of the injection front. Within the oil bank, only oil flows, at constant saturation (equal to $1 - S_{wi}$); ahead of the bank, only gas flows (at saturation S_{gi}), and behind it, only water does (at saturation $1 - S_{orw}$).
7. A residual oil saturation, S_{orw}, is left behind after all the gas and movable oil have been displaced, and the remaining pore space is occupied by water at saturation $1 - S_{orw}$.

Hence, at the outflow boundary (the producer), gas will be produced, with no liquid, until all the free gas has been displaced and the leading edge of the oil bank reaches the producer. Then, oil only will be produced until all the movable oil has been displaced (*i.e.*, the trailing edge of the oil bank reaches the producer), and, finally, water only is produced until injection stops.

Consider a system of length L and cross-sectional area A perpendicular to the direction of flow, and let V_p, V_{fu}, and V_{fo} represent, respectively, its total pore volume, fill-up volume, and flood-out volume:

$$V_p = \phi \cdot A \cdot L \tag{1}$$

$$V_{fu} = S_{gi} \cdot V_p \tag{2}$$

$$V_{fo} = (S_{gi} + S_{oi} - S_{orw}) \cdot V_p \tag{3}$$

Let i_w be the rate of water injection, q_o and q_w be the oil and water volumetric production rates, respectively, and W, Q_o, and Q_w be the cumulative water injected, oil produced, and water produced, respectively. Then

$$W = i_w t \tag{4}$$

$$
\begin{aligned}
Q_o &= 0, & &\text{for } W < V_{fu}; \\
&= W - V_{fu}, & &\text{for } V_{fu} \leq W < V_{fo}; \\
&= V_{fo} - V_{fu}, & &\text{for } V_{fo} \leq W
\end{aligned}
\tag{5}
$$

$$
\begin{aligned}
Q_w &= 0, & &\text{for } W < V_{fo}; \\
&= W - V_{fo}, & &\text{for } V_{fo} \leq W.
\end{aligned}
\tag{6}
$$

Note that the limiting value of cumulative oil production, Q_o, is equal to the difference between the flood-out volume and the fill-up volume, $V_{fo} - V_{fu}$, which is just the total movable oil in place at the start of waterflooding, V_{mo}.

The production rates, q_o and q_w, can be obtained simply by differentiation of the equations for Q_o and Q_w:

$$q_o = 0, \qquad \text{for } W < V_{fu};$$
$$= i_w, \qquad \text{for } V_{fu} \leq W < V_{fo};$$
$$= 0, \qquad \text{for } V_{fo} \leq W \qquad\qquad (7)$$
$$q_w = 0, \qquad \text{for } W < V_{fo};$$
$$= i_w, \qquad \text{for } V_{fo} \leq W. \qquad\qquad (8)$$

It is convenient to normalize equations 5 and 6, and the associated inequalities, by dividing both sides by V_{mo}, letting $\gamma = W/V_{mo}$ be the normalized cumulative injection — i.e., the volume of water injected per unit volume of movable oil in place at the start of waterflooding:

$$Q_o^* = 0, \qquad \text{for } \gamma < \lambda;$$
$$= \gamma - \lambda, \qquad \text{for } \lambda \leq \gamma < 1 + \lambda,$$
$$= 1, \qquad \text{for } 1 + \lambda \leq \gamma. \qquad\qquad (9)$$
$$Q_w^* = 0, \qquad \text{for } \gamma < 1 + \lambda,$$
$$= \gamma - 1 - \lambda, \qquad \text{for } 1 + \lambda \leq \gamma. \qquad\qquad (10)$$

Here, Q_o^* and Q_w^* are the normalized cumulative oil and water production, respectively. Note that the former quantity is just the oil recovery efficiency, expressed as a fraction of the movable oil in place at the start of waterflooding. The parameter λ is the ratio of fill-up volume to movable oil volume. No oil is produced until the normalized cumulative injection is equal to λ. The recovery efficiency then increases linearly in direct proportion to normalized injection until all the oil has been produced (or injection stops).

Equations 7 and 8 are conveniently normalized by dividing the oil and water production rates by the water injection rate; the inequalities are normalized as before by dividing them through by the movable oil volume:

$$q_o^* = 0, \qquad \text{for } \gamma < \lambda;$$
$$= 1, \qquad \text{for } \lambda \leq \gamma < 1 + \lambda,$$
$$= 0, \qquad \text{for } 1 + \lambda \leq \gamma. \qquad\qquad (11)$$

$$q_w^* = 0, \qquad \text{for } \gamma < 1 + \lambda,$$
$$= 1, \qquad \text{for } 1 + \lambda \leq \gamma. \qquad\qquad (12)$$

Figure 2 shows a plot of the normalized oil production rate vs. normalized cumulative injection as given by equation 11. Plots for the other normalized variables defined in equations 9, 10, and 12 are not shown, because they can easily be sketched by the reader.

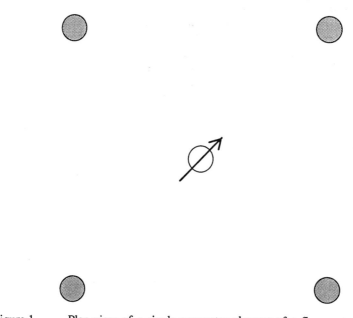

Figure 1.   Plan view of a single symmetry element of a five-spot waterflood pattern. The solid circles represent the producing wells, and the circle with the arrow through it represents the injection well.

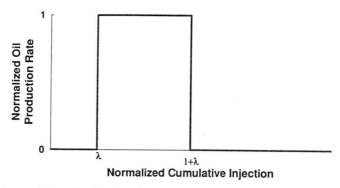

Figure 2.   Normalized oil production rate vs. normalized cumulative injection for an ideal homogeneous waterflood

Waterflood Efficiency in A Heterogeneous System

With the ideal linear system as a starting point, we can now consider the effect of the spatial variation in permeability. In order to simplify the analysis, the heterogeneous system is represented conceptually as a collection of independent linear elements connecting the inflow boundary (or boundaries) with the outflow boundary (or boundaries). The jth element has intrinsic permeability k_j and area a_j perpendicular to the direction of flow, and all elements are assumed to have the same length, porosity, and fluid saturations. The total area perpendicular to the direction of flow is A.

By ignoring changes in flow resistance (i.e., neglecting the saturation-dependence of relative permeabilities) as the displacement fronts propagate through each element, and recognizing that the pressure drop between inflow and outflow must be the same for each element, we can write the cumulative injection into the jth layer, W_j, as:

$$W_j = \frac{k_j \cdot a_j}{\sum\limits_{J=1}^{N} k_j \cdot a_j} \cdot W$$

$$= \frac{k_j \cdot a_j}{\overline{k} \cdot A} \cdot W \tag{13}$$

Upon multiplying the numerator and denominator by $\phi \cdot L \cdot \left(S_{oi} - S_{orw} \right)$ and rearranging the result, the normalized injection variable for element j is found to be:

$$\gamma_j = \gamma \cdot k_j / \overline{k} \tag{14}$$

when the normalized cumulative injection into the entire system is equal to γ. Each element will obey equations 7 and 8, with γ_j in place of γ:

$$Q_o^{\bullet}\left(k_j, \gamma \right) = 0, \qquad \text{for } k_j < \frac{\lambda \cdot \overline{k}}{\gamma} \equiv k_1;$$

$$= \frac{k_j \cdot \gamma}{\overline{k}} - \lambda, \quad \text{for } k_1 \leq k_j < \frac{(1+\lambda) \cdot \overline{k}}{\gamma} \equiv k_2,$$

$$= 1, \qquad \text{for } k_2 \leq k_j. \tag{15}$$

$$Q_w^{\bullet}\left(k_j, \gamma \right) = 0, \qquad \text{for } k_j < k_2,$$

$$= \frac{k_j \cdot \gamma}{\overline{k}} - 1 - \lambda, \quad \text{for } k_2 \leq k_j. \tag{16}$$

Note that in equations 15 and 16, the cumulative normalized oil and water production are written as functions of k_j and γ, and that the associated inequalities are expressed as inequalities for k_j at a given value of total dimensionless injection, γ. At a given value of γ, all elements with k less than k_1 will have produced no oil, those with k between k_1 and k_2 will have produced only a fraction of their oil, and those with k

greater than k_2 will have produced all of their oil. Similarly, elements with k less than k_2 will have produced no water, while those with k greater than k_2 continue to produce water so long as it is injected.

To obtain the total normalized production of oil and water, we must sum the contributions of all the elements, weighted by their volume fractions in the total system:

$$Q_o^{\bullet}(\gamma) = \sum_{j\ni k_j < k_1} 0 \cdot \frac{a_j}{A} + \sum_{j\ni k_1 \le k_j < k_2} \left[\frac{k_j \cdot \gamma}{\overline{k}} - \lambda \right] \cdot \frac{a_j}{A} + \sum_{j\ni k_2 \le k_j} 1 \cdot \frac{a_j}{A} \qquad (17)$$

$$Q_w^{\bullet}(\gamma) = \sum_{j\ni k_j < k_2} 0 \cdot \frac{a_j}{A} + \sum_{j\ni k_2 \le k_j} \left[\frac{k_j \cdot \gamma}{\overline{k}} - 1 - \lambda \right] \cdot \frac{a_j}{A} \qquad (18)$$

With one additional assumption, the volumetric averages in equations 17 and 18 can be replaced by averages over the *permeability distribution* of the system. We assume that this distribution is described by a probability density function $f(k)$, and that $f(k)dk$ represents the volume fraction with permeability between k and $k + dk$, in accordance with the usual frequency interpretation of probability density functions. Hence, we can replace the fraction a_j/A in the above equations by $f(k)dk$, and, in the limit as dk approaches zero, the sums become integrals:

$$Q_o^{\bullet}(\gamma) = \frac{\gamma}{\overline{k}} \int_{k_1}^{k_2} kf(k)dk - \lambda \int_{k_1}^{k_2} f(k)dk + \int_{k_2}^{\infty} f(k)dk \qquad (19)$$

$$Q_w^{\bullet}(\gamma) = \frac{\gamma}{\overline{k}} \int_{k_2}^{\infty} kf(k)dk - (1+\lambda)\int_{k_2}^{\infty} f(k)dk \qquad (20)$$

For many aquifers and petroleum reservoirs, the distribution of permeability values measured on cores is closely approximated by a log-normal distribution, with density function

$$f(k) = \frac{\exp\left[-\frac{1}{2}\left(\frac{\ln k - \mu}{\sigma} \right)^2 \right]}{k\sigma\sqrt{2\pi}} \qquad (21)$$

The parameters μ and σ are, respectively, the mean of the natural logarithm of k and the square root of the variance of the natural logarithm of k. If we assume that the distribution of permeability is log-normal in equations 19 and 20, the integrations can be performed explicitly.

Note that there are only two types of integrals over the log-normal density function which are needed: one involving a partial average of k, and one involving an incomplete integral of the density function. By some straightforward but tedious manipulation involving changing the variable of integration and integrating by parts, it can be shown that, for arbitrary values of a and b:

$$\int_{a\bar{k}}^{b\bar{k}} f(k)dk = \Phi\left[\frac{\ln(b)+\dfrac{\sigma^2}{2}}{\sigma}\right] - \Phi\left[\frac{\ln(a)+\dfrac{\sigma^2}{2}}{\sigma}\right] \tag{22}$$

$$\int_{a\bar{k}}^{b\bar{k}} kf(k)dk = \bar{k}\left\{\Phi\left[\frac{\ln(b)-\dfrac{\sigma^2}{2}}{\sigma}\right] - \Phi\left[\frac{\ln(a)-\dfrac{\sigma^2}{2}}{\sigma}\right]\right\} \tag{23}$$

$$\Phi(z) = \frac{1}{\sqrt{2\pi}}\int_{-\infty}^{z} e^{-x^2/2} \text{ is the normal probabili ty integral,} \tag{24}$$

resulting in the following expressions for dimensionless cumulative oil and water production, respectively:

$$Q_o^{\bullet} = \gamma\left\{\Phi\left[\frac{\ln\left(\dfrac{\gamma}{\lambda}\right)+\dfrac{\sigma^2}{2}}{\sigma}\right] - \Phi\left[\frac{\ln\left(\dfrac{\gamma}{1+\lambda}\right)+\dfrac{\sigma^2}{2}}{\sigma}\right]\right\}$$

$$+(1+\lambda)\Phi\left[\frac{\ln\left(\dfrac{\gamma}{1+\lambda}\right)-\dfrac{\sigma^2}{2}}{\sigma}\right] - \lambda\Phi\left[\frac{\ln\left(\dfrac{\gamma}{\lambda}\right)-\dfrac{\sigma^2}{2}}{\sigma}\right] \tag{25}$$

$$Q_w^{\bullet} = \gamma\Phi\left[\frac{\ln\left(\dfrac{\gamma}{1+\lambda}\right)+\dfrac{\sigma^2}{2}}{\sigma}\right] - (1+\lambda)\Phi\left[\frac{\ln\left(\dfrac{\gamma}{1+\lambda}\right)-\dfrac{\sigma^2}{2}}{\sigma}\right] \tag{26}$$

Normalized rates of production can be obtained by differentiating equations 25 and 26 with respect to γ, and noting that

$$\frac{q_o}{i_w} = \frac{dQ_o^{\bullet}}{d\gamma} \tag{27}$$

and

$$\frac{q_w}{i_w} = \frac{dQ_w^{\bullet}}{d\gamma} \tag{28}$$

Again, the algebra is tedious but straightforward, and the resulting expressions are:

$$\frac{q_o}{i_w} = \Phi\left[\frac{\ln\left(\frac{\gamma}{\lambda}\right) + \frac{\sigma^2}{2}}{\sigma}\right] - \Phi\left[\frac{\ln\left(\frac{\gamma}{1+\lambda}\right) + \frac{\sigma^2}{2}}{\sigma}\right] \tag{29}$$

$$\frac{q_w}{i_w} = \Phi\left[\frac{\ln\left(\frac{\gamma}{1+\lambda}\right) + \frac{\sigma^2}{2}}{\sigma}\right] \tag{30}$$

In obtaining the final forms of equations 25, 26, 29, and 30, the fact that $\Phi(-z) = 1 - \Phi(z)$ has been used.

Note that these expressions for dimensionless cumulative oil and water production and oil and water production rates reduce to those given previously for the ideal homogeneous case when σ approaches 0. For $\sigma = 0$, the normalized oil rate is a rectangular pulse, equal to zero until $\gamma = \lambda$, when it jumps to 1.0, and remains there until $\gamma = 1 + \lambda$, when it returns to zero. This is precisely the behavior shown in Figure 2. The normalized water rate is a unit step function at $1+\lambda$. In fact, the oil pulse for the homogeneous case is just the difference between a step function for the leading edge of the oil bank and a step function for the trailing edge of the oil bank.

The median for each log-normal function in the heterogeneous system is the corresponding step-function argument multiplied by $\exp(-\sigma^2/2)$. Hence, as the heterogeneity parameter increases, the median breakthrough of each displacement front occurs at smaller and smaller values of cumulative injection.

The derivation of the waterflood model has been presented in some detail since it provides a comprehensible physical basis for the introduction of heterogeneity as a log-normal transformation on a step function representing the propagation of a displacement front in an ideal homogeneous system.

A Field Example of a Waterflood

There are several consequences of this simplified treatment that have practical significance. One is that for displacements in heterogeneous systems, the relevant operating variable is the *logarithm* of cumulative injection (or time, for more-or-less constant injection rates), which very vividly illustrates the rapidly diminishing return for any recovery (or remediation) process involving fluid displacement. For the specific case of a waterflood, a plot of oil production rate vs. time on a logarithmic scale should result in a symmetric Gaussian curve shape, as shown in Figure 3 for an Illinois Basin waterflood (Benton Field) from 1947 until 1972.

Perhaps even more significant is that the size of the system does not appear explicitly in the normalized response equations. Changing the total movable oil in place merely translates the model production response curve along the horizontal (logarithmic) axis. If the vertical axis is made logarithmic as well, changing scales translates the curve horizontally and vertically without changing its shape. This implies that, provided there is some statistical regularity within a single reservoir being waterflooded, the shape of production curves for individual wells or groups of wells should be similar to the curve for the entire reservoir when plotted on log-log axes. This conclusion was shown to be correct, for the Benton waterflood, in reference 1.

Finally, there is a self-contained test of the theory. Note that the normalized water production rate is given by a single log-normal probability integral (equation 30) and that the sum of oil and water (*i.e.*, total liquid) production rates is given by a different log-normal integral with the same σ but a different median. Plots of normalized water and total liquid production rates vs. log of time on probability paper should result in parallel straight lines.

Figures 4 and 5 show, respectively, probability plots of total liquid production rate vs. log W and water production rate vs. log W. In addition to the field data, these plots also show straight lines obtained from a least-squares fit to the data. The slopes of these lines give directly the values of σ from the total liquid response (0.813) and from the water production response (0.804). If the model is a good approximation to the displacement of fluids in heterogeneous media, these σ values should be identical. The agreement is quite good, especially in view of the gross simplifying assumptions made in deriving the model equations, and suggests that the shape of breakthrough curves for field-scale transport processes may be dominated by spatial heterogeneities rather than by classical dispersion theory, which predicts a Gaussian breakthrough curve. More examples and details of fitting field data for waterfloods are given in reference 1.

Vapor Extraction

Ideal System. To develop a model for vapor extraction, we again consider a linear element with area A perpendicular to the direction of flow, and let S_v be the total vapor content (air plus water vapor plus contaminant plus soil gas) per unit pore volume. The total bulk volume of sediment from which gas can flow to the outflow boundary (*e.g.*, an extraction well) is assumed to be divided into two regions: an uncontaminated volume, V_u, nearest the outflow, and a contaminated volume, V_c, further from the outflow. This division is included to allow for a delay between the start of gas extraction and the beginning of contaminant removal — *i.e.*, to allow for the case in which the extraction well is not located within the contaminant plume.

Within the contaminated volume, a volatile contaminant is assumed to be present in the vapor phase at concentration C_{vo} (mass per unit volume of vapor phase), in equilibrium with contaminant at concentration C_{lo} (mass per unit volume of aqueous phase) dissolved in a stationary aqueous phase held by capillary forces in the porous medium.

Water saturation is assumed equal to $1-S_v$; in other words, there is no free non-aqueous phase liquid (NAPL) present. Solid-phase partitioning is also ignored, although it could be included (3). Vapor is extracted at a constant volumetric rate q_v at the outflow well, and is supplied at some effective outer boundary either by injection, natural influx, or a combination of the two. Gas-phase compressibility is ignored.

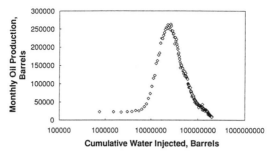

Figure 3.   Monthly Oil Production versus Cumulative Water Injected for the Benton Waterflood, Benton, Illinois.

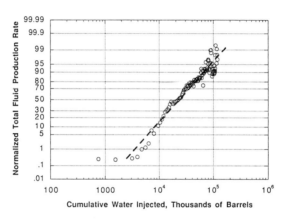

Figure 4.   Log-normal probability plot of normalized total fluid production rate vs. cumulative water injected, for field data (circles) and a linear least-squares fit to the data (solid line).

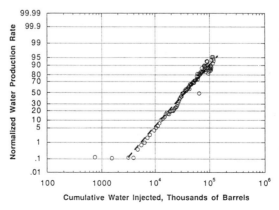

Figure 5.   Log-normal probability plot of normalized water production rate vs. cumulative water injected, for field data (circles) and a linear least-squares fit to the data (solid line).

Consider a location x at time t between the inflow boundary at $x = 0$ and the outflow boundary at $x = L$. At distances less than x, the medium is contaminated, and at greater distances the medium is uncontaminated. In other words, x is the position of the contaminant front (assumed sharp) at time t. In time Δt, the carrier fluid (air) will move from x to $x + \Delta x$, where

$$\Delta x = \frac{q_v}{\phi S_v A} \Delta t. \tag{31}$$

Contaminant movement is retarded by partitioning into the stationary liquid phase. Let f be the fraction of the length Δx which becomes contaminated in time Δt — i.e., the contaminant front will advance a distance $f \Delta x$.

A mass balance on the contaminant gives

$$C_{vo} q_v \Delta t = f \Delta x A \phi S_v C_{vo} + f \Delta x A \phi (1 - S_v) C_{lo}$$

$$= f\left(\frac{q_v}{\phi S_v A} \Delta t\right) A \phi S_v C_{vo} + f\left(\frac{q_v}{\phi S_v A} \Delta t\right) A \phi (1 - S_v) C_{lo}, \tag{32}$$

where equation 31 has been used to substitute for Δx. After canceling common terms, the resulting expression can be solved for f:

$$f = \left(1 + \left[\frac{1 - S_v}{Sv}\right]\frac{C_{lo}}{C_{vo}}\right)^{-1} \tag{33}$$

The velocity of the contaminant front is

$$v_{cf} = f\frac{\Delta x}{\Delta t} = \frac{q_v}{\phi S_v A}\left(1 + \left[\frac{1 - S_v}{Sv}\right]\frac{C_{lo}}{C_{vo}}\right)^{-1}, \tag{34}$$

where equation 31 was used to obtain the carrier fluid velocity. A similar analysis shows that the trailing edge of the contaminant moves with the same velocity. Hence, the contaminant moves as a rectangular pulse in this idealized model, exactly analogous to the movement of the oil bank in a waterflood. To complete the analogy, we need to determine the cumulative carrier fluid (air) throughput at the time the leading edge of the contaminant pulse reaches the outflow boundary, and the cumulative throughput when the trailing edge reaches the boundary.

Let the region from $x=0$ to $x = L_c$ be the originally contaminated portion, and the region from L_c to L be the uncontaminated portion of the system. Let t_f be the time required for the front of the contaminant pulse to reach the outflow, and t_b be the time for the back to reach the outflow. These are readily obtained by using equation 34 for the contaminant pulse velocity, and noting that the distances traveled are $L-L_c$ and L, respectively:

$$t_f = \frac{L - L_c}{v_{cf}} = \frac{L_c}{v_{cf}}(L / L_c - 1) \equiv \frac{\lambda L_c}{v_{cf}} \tag{35}$$

$$t_b = \frac{L}{v_{cf}} = \frac{L_c}{v_{cf}}(L / L_c) \equiv \frac{(1+\lambda)L_c}{v_{cf}} \qquad (36)$$

where the parameter λ has been introduced to maintain the analogy with the waterflood model. Note that it has the same effect in the ideal contaminant transport model that it does in the ideal waterflood model: *viz.*, it introduces a delay in contaminant breakthrough at the outflow boundary of the system.

The form of equations 35 and 36 suggests that an appropriate normalized cumulative throughput variable for vapor extraction can be defined as:

$$\gamma_v = \frac{v_{cf}t}{L_c} \qquad (37)$$

since contaminant extraction starts when γ_v is equal to λ (*i.e.*, $t = t_f$) and ends when it is equal to $1+\lambda$ (*i.e.*, $t = t_b$).

A more physically transparent definition for γ_v is obtained by substituting the right hand side of equation 34 for v_{cf}:

$$\frac{v_{cf}}{L_c} = \frac{q_v}{L_c \phi S_v A \left(1 + \left[\frac{1-S_v}{S_v} \right] \frac{C_{lo}}{C_{vo}} \right)}$$

$$= \frac{C_{vo}q_v}{M_{co}} \qquad (38)$$

Note that the total initial mass of contaminant in the system, M_{CO} is given by:

$$M_{co} = L_c A \left[C_{vo} \phi S_v + C_{lo} \phi (1 - S_v) \right]$$

$$= L_c A \frac{C_{vo} \phi S_v}{f} \qquad (39)$$

Finally, the normalized throughput can be written as:

$$\gamma_v = \left(\frac{C_{vo}}{M_{co}} \right) (q_v t)$$

$$\equiv \frac{W_v}{(M_{co} / C_{vo})} \qquad (40)$$

where the numerator is the cumulative volume of vapor (air, soil gas, water and contaminant vapor) removed at the outflow boundary, and the denominator is the amount that would have to be removed to extract all of the initial contaminant mass if λ were equal to zero and the entire contaminant inventory could be extracted at constant concentration C_{vo}. This definition for normalized throughput hence has an analogous physical significance to the corresponding normalized quantity for a waterflood.

In complete analogy with equations 9 and 11 for the normalized cumulative oil production and the normalized oil production rate for an ideal waterflood, we can write the following equations for normalized cumulative contaminant removal and normalized contaminant concentration in the produced gas for an ideal vacuum extraction process:

$$M / M_{co} = 0, \qquad \text{for } \gamma_v < \lambda;$$

$$= \gamma_v - \lambda, \qquad \text{for } \lambda \le \gamma_v < 1 + \lambda,$$

$$= 1, \qquad \text{for } 1 + \lambda \le \gamma_v. \qquad (41)$$

$$C_v / C_{vo} = 0, \qquad \text{for } \gamma_v < \lambda;$$

$$= 1, \qquad \text{for } \lambda \le \gamma_v < 1 + \lambda,$$

$$= 0, \qquad \text{for } 1 + \lambda \le \gamma_v. \qquad (42)$$

Here, M is the cumulative total contaminant mass removed as a function of the normalized cumulative total volume of gas extracted, and C_V is the corresponding contaminant concentration in the extracted gas.

Heterogeneous System.   We can immediately write down the equations for a heterogeneous system comprised of infinitesimal linear elements with a log-normal distribution of permeability. The detailed steps are exactly the same as those used in deriving the waterflood equations. For the cumulative normalized contaminant mass extracted, we have

$$M / M_{co} = \gamma_v \left\{ \Phi \left[\frac{\ln\left(\dfrac{\gamma_v}{\lambda}\right) + \dfrac{\sigma^2}{2}}{\sigma} \right] - \Phi \left[\frac{\ln\left(\dfrac{\gamma_v}{1 + \lambda}\right) + \dfrac{\sigma^2}{2}}{\sigma} \right] \right\}$$

$$+ (1 + \lambda) \Phi \left[\frac{\ln\left(\dfrac{\gamma_v}{1 + \lambda}\right) - \dfrac{\sigma^2}{2}}{\sigma} \right] - \lambda \Phi \left[\frac{\ln\left(\dfrac{\gamma_v}{\lambda}\right) - \dfrac{\sigma^2}{2}}{\sigma} \right] \qquad (43)$$

and for the normalized contaminant concentration in the extracted gas, we have

$$C_v / C_{vo} = \Phi \left[\frac{\ln\left(\dfrac{\gamma_v}{\lambda}\right) + \dfrac{\sigma^2}{2}}{\sigma} \right] - \Phi \left[\frac{\ln\left(\dfrac{\gamma_v}{1 + \lambda}\right) + \dfrac{\sigma^2}{2}}{\sigma} \right] \qquad (44)$$

Equations 43 and 44 reduce to equations 41 and 42, respectively, as σ approaches zero.

To illustrate the effect of the heterogeneity parameter, σ, on the cumulative contaminant extraction and the concentration of contaminant in the extracted gas,

Figure 6 was prepared, assuming no delay between the start of vapor extraction and the arrival of contaminant at the extraction well (*i.e.,* $\lambda = 0$). Note that a log-log scale is used in Figure 6. As pointed out in the discussion of the analogous waterflood model, the shapes of these curves should be invariant with respect to changes in scale factors when plotted with log-log axes. Hence the parameters λ and σ can be determined by plotting field data on log-log paper, and comparing the resulting plots with a family of type curves for the normalized responses with systematically varying values of λ and σ. The translations required along the horizontal and vertical axes to achieve a match between a type curve and the field plot can then be used to determine the parameters C_{vo} and M_{co} by selecting a match point and reading off the corresponding values of the dimensionless quantities and field variables.

It is perhaps noteworthy that the general effects shown by these curves are consistent with observed performance in vapor extraction projects. In the ideal homogeneous linear system, contaminant is removed at a constant rate until it is all extracted, as shown in Figure 6 for $\sigma = 0$. Of course, this ideal behavior is never obtained in the field; in reality, the contaminant concentration in the extracted vapor decreases rapidly at first, but then continues to be produced at lower concentrations for a very long time. These lower concentrations remain significantly higher than regulatory standards until many times the ideal volume of gas has been extracted. This behavior is shown by the curves in Figure 6 for $\sigma \neq 0$.

For the examples shown in Figure 6, note that attaining the three - to - four order of magnitude reduction in contaminant concentration typically needed to meet regulatory standards requires on the order of 10 times the ideal throughput volume for a system as heterogeneous as the Benton waterflood. For even more heterogeneous systems, 100, 1000, or even larger multiples of the ideal throughput volume would be required to achieve regulatory compliance, depending upon the effective value of σ.

Savannah River In-Situ Air Stripping Test

At the Savannah River Plant, the subsurface below A-Area and M-Area is contaminated by volatile organic compounds (VOCs), principally Trichloroethylene (C_2HCl_3, or TCE) and Tetrachloroethylene (C_2Cl_4, or PCE). These compounds were used extensively as degreasing solvents from 1952 until 1979, and the waste solvent which did not evaporate (on the order of 2×10^6 pounds) was discharged to a process sewer line leading to the M-Area Seepage Basin. These compounds infiltrated into the soil and underlying sediments from leaks in the sewer line and elsewhere, thereby contaminating the vadose zone between the surface and the water table as well as the aquifer.

As part of the Integrated Demonstrations for the development of new remediation technologies by the U.S. Department of Energy (DOE), Westinghouse Savannah River Corporation initiated a project to drill and install two horizontal wells, AMH-1 and AMH-2, for testing of vadose-zone remediation processes (*4* and *5*). These wells were drilled and completed in September and October, 1988, and were used in the Air-Stripping Phase of the Savannah River Integrated Demonstration Project for the Removal of VOCs at Non-Arid Sites from July 7 to December 13, 1990.

The extraction well, AMH-2, has a screened interval of approximately 200 feet entirely within the vadose zone, and the injection well, AMH-1, has a screened interval below the water table of about 310 feet (6).

Vacuum extraction at approximately 580 SCFM from AMH-2 started July 27, 1990 and continued, with minor interruptions, until December 13, 1990 (7). During part of the test period, air was injected at various rates below the water table through AMH-1, to test the possibility of air-stripping VOCs from groundwater *in situ* while using vacuum extraction to remove contaminants from the vadose zone.

During the first 21 days of operation, a total of approximately 2696 lb. of VOCs was removed. For the first 15 days, only the extraction well was used. Injection into AMH-1 at 65 SCFM began on day 16, and the rate was increased to 140 SCFM on day 28 and to 270 SCFM on day 69. Injection stopped on day 113, and extraction continued to day 140. These periods are elapsed time from the start of the test. Approximately 120 days of actual operating time were achieved, accomplishing the removal of about 15900 pounds of VOCs.

Application of the Heterogeneity Parameter Model

As a preliminary step in applying the model equations for vapor extraction in heterogeneous systems, concentrations of TCE and PCE in the gas extracted from AMH-2 were plotted versus hours of net operating time (*i.e.*, with down time subtracted), using a log-log scale, in Figure 7. The resulting curves bear only a slight resemblance to the normalized model curves previously shown in Figure 6.

One reason for this is that all of the Figure 6 curves are for $\lambda = 0$. It can be shown that the normalized concentration predicted by the model (equation 44) has a relative maximum at a normalized extraction value γ^* given by

$$\gamma^* = e^{-\sigma^2/2} \sqrt{\lambda(1+\lambda)} \qquad (45)$$

The maximum value of the normalized concentration is

$$\left(\frac{C_v}{C_{vo}}\right)_{max} = 2\Phi\left[\frac{\ln\left(\dfrac{1+\lambda}{\lambda}\right)}{2\sigma}\right] - 1 \qquad (46)$$

For $\lambda = 0$, equation 45 shows that the maximum concentration occurs at the start of extraction (time or cumulative extraction equal to zero) and equation 46 shows that the maximum value of the normalized concentration is equal to unity. As the delay parameter λ increases, the maximum normalized concentration decreases rapidly. The curves in Figure 7 are consistent with a local maximum even if the two earliest points are ignored.

Another complication arises from the change in operating conditions during the course of the test. Many simplifying assumptions were made in the derivation of the model equations. One of the more important of these is the implicit assumption that the pore volume affected by the extraction well remains constant over time. This is probably justified if operating conditions are held constant and the flow properties of the system are such that a pseudo-steady state is reached in a time short compared to the duration of the operation.

There were several significant changes during the operation of the In-Situ Air Stripping Test. Injection at 65 SCFM was started 342.5 operating hours after extraction began, was increased to 170 SCFM at 655.0 hours, and then increased again to 270 SCFM at 1511.6 hours. Injection was terminated at 2261.6 hours, and extraction continued until 2879.1 hours. These changes should alter the capture volume of AMH-1.

To attempt fitting the data, it was decided to concentrate first on the initial 342.5 hours, when the operation was a pure vapor extraction process. A mathematical spreadsheet was developed (using commercially available software) to import the field concentration vs. time data, calculate the normalized concentration vs. normalized cumulative extraction from equation 44 for any desired combination of the model parameters, and graph both the field data and model results on the same log-log plot so they can be compared visually.

Examination of the model equations shows that there are four parameters to be determined: λ, σ, C_0, and C_1. The first two are dimensionless parameters which affect the *shape* of the concentration response curve when plotted on log-log paper. The second pair are scale factors which translate the log-log response curve parallel to the vertical and horizontal axis, respectively, without changing its shape.

Since the field concentration data are given in ppm by volume, and any consistent units can be used for the normalized concentration appearing in the left hand side of equation 44, it is convenient to choose the vertical scale factor C_0 to match the field data as reported, then convert to mass per unit volume as follows:

$$C_{vo} = 10^{-6} C_0 \left(\frac{MW}{359.1} \right) \qquad (47)$$

In equation 47, C_0 is in ppm by volume, MW is the molecular weight in lb.$_m$ per lb.-mole, and the constant in the denominator is the number of standard cubic feet per lb.-mole of an ideal gas. The concentration C_{vo} will then be in lb.$_m$ per SCF.

The parameter C_1 for translation along the horizontal axis maps the normalized cumulative extraction into a corresponding value for the independent throughput variable. Since the extraction rate was treated as a constant in deriving the model equations, either time or cumulative extraction can be used as an independent variable for plotting the response curve. Time is somewhat more convenient if the injection rate does not vary much. If it does vary significantly, it will often be advantageous to use the cumulative extraction as the horizontal scale. We chose to use time because the extraction rate did not vary much except when the system was down. Hence we seek a scale factor by matching the plots such that

$$t = C_1 \gamma \qquad (48)$$

Since γ is dimensionless, it is obvious that C_1 is a characteristic time. From equation 40, it is easy to show that

$$C_1 = \frac{M_{co}}{q_v C_{vo}} \tag{49}$$

After determining values for the two scale factors C_0 and C_1 by matching the field data, and calculating C_{vo} from equation 47, the total initial mass inventory can calculated from

$$M_{co} = C_{vo} C_1 q_v \tag{50}$$

In using equation 50, care must be taken to use consistent units. For time in hours and concentration in lb.$_m$ per cu. ft., the volumetric extraction rate must be in cubic feet per hour. The mass inventory will then be obtained in pounds mass.

Figure 8 shows a plot from the mathematical spreadsheet for fitting the model to the early time data for TCE concentration in the extracted gas. The agreement between the model and field data is visually quite good before air injection started. There is an increasing spread between the model and the field results as the operation continues, indicating that air injection may be sweeping contaminant to the extraction well which would not be extracted otherwise. No attempt was made to match the cumulative contaminant mass withdrawal directly.

Figure 9 shows the match for the early-time PCE data. Again, the match is quite good until about the time that air injection began, with an even more pronounced departure afterward.

The parameters from the early-time match were used in a second spreadsheet to calculate concentrations of TCE and PCE in ppm for each time they were measured, and the calculated values were subtracted from observed values to obtain residuals. These residuals were *assumed* to result from the sweep of additional contaminant inventory to the extraction well as a result of injection, and the effect of injection *rate* was ignored.

The spreadsheet was then used to fit these residual concentrations with a second set of parameter values for each contaminant. The results for TCE are shown in Figure 10; those for PCE are shown in Figure 11. Although there is considerable scatter in the data, the overall agreement looks reasonable. More emphasis was given to obtaining a match at later time than at early time, since the injection conditions were held constant for a longer interval. Parameter values for the early- and late-time models are given in Table I for both TCE and PCE.

Table I. Summary of heterogeneity model parameters obtained by fitting field data for the concentration of VOCs in the gas extracted during the Savannah River Integrated Demonstration of In-Situ Air Stripping. The total TCE inventory is estimated to be about 21,000 lb., and the PCE inventory estimate is just below 60,000 lb.

Parameter	$TCE_{(VE)}$	$TCE_{(AS)}$	$PCE_{(VE)}$	$PCE_{(AS)}$
λ	0.10	8.00	0.01	0.5
σ	2.5	0.8	2.5	1.1
C_0	1000	1700	1500	700
C_1	800	500	150	5000
M_{co}	10186	10823	3616	56246

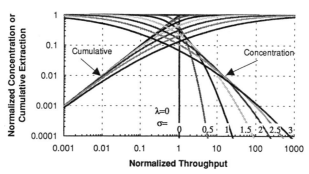

Figure 6.　Normalized contaminant concentration and normalized cumulative contaminant mass removed vs. normalized cumulative total vapor extracted (throughput) for $\lambda = 0$, for values of σ ranging from 0 to 3. Note that there is no theoretical upper limit to the value of σ

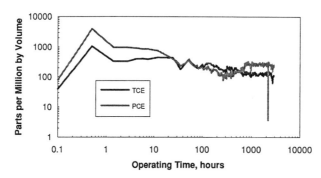

Figure 7.　Contaminant concentrations (ppm by volume) measured in gases extracted from horizontal well AMH-2 during the Savannah River Integrated Demonstration of In-Situ Air Stripping.

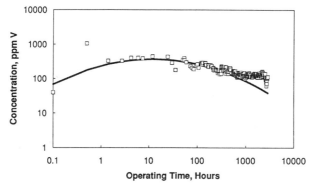

Figure 8.　Fit of early-time TCE extraction data to the heterogeneity parameter model. The squares are the data points, and the smooth curve is calculated from equations 44 and 48 with $\lambda = 0.10$, $\sigma = 2.5$, $C_0 = 1000$, and $C_1 = 800$.

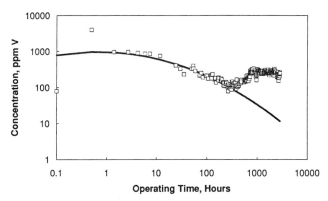

Figure 9.   Fit of early-time PCE extraction data to the heterogeneity parameter model. The squares are the data points, and the smooth curve is calculated from equations 44 and 48 with $\lambda = 0.01$, $\sigma = 2.5$, $C_0 = 1500$, and $C_1 = 150$.

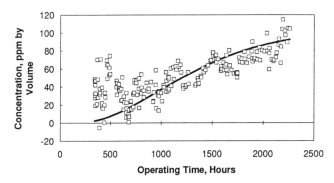

Figure 10.   Residual TCE concentrations compared to model calculations with parameters chosen to fit the residual data. The parameter values are shown in Table I.

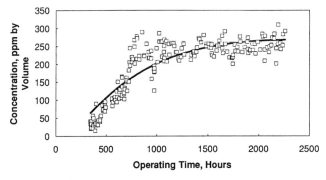

Figure 11.   Residual PCE concentrations compared to model calculations with parameters chosen to fit the residual data. The parameter values are shown in Table I.

The spreadsheet was then used to calculate concentrations vs. time for each set of parameters for each contaminant, and the resulting pairs of concentration values for TCE and PCE were added together at each time to produce total concentration vs. time curves for each contaminant. The results are shown and compared with field data in Fig 12. The agreement is quite good.

Finally, equation 43 was used to calculate the cumulative mass extracted for each component for both parameter sets, and the results for each separate component were added together. The resulting cumulative extraction plots are given in Figures 13 and 14. Cumulative extraction estimates from the measured concentration data and the nominal extraction rate are 5218 lb. for TCE and 11149 lb. for PCE. Comparable model results are 5323 lb. and 11511 lb., respectively. The model estimate is about 2% above actual for TCE and about 3% above actual for PCE, which is remarkably close agreement. However, it should be quickly noted that there are a number of parameters to adjust, and this agreement may merely reflect the existence of enough degrees of freedom to match almost any data set. If we tentatively accept the total mass inventory indicated by the model parameters chosen to fit the data, the VOC recovery of about 16000 lb. is only about 20% of the indicated initial contaminant inventory of approximately 81000 lb.

Conclusion

Before discussing the results presented above for the use of this simple analytical heterogeneity parameter model to evaluate the Savannah River Integrated Demonstration, I would like to emphasize that this paper should be regarded as a progress report in the development of an approach to quantifying the effects of field-scale heterogeneities on remediation processes. It is not a finished product, and much work remains to be done even with Savannah River data before taking the results too seriously. However, the general approach has worked well when applied to waterflooding reservoirs with enough core data to determine accurate *a priori* estimates for the model parameters (pore volumes, saturations, and the variance of intrinsic permeability), provided that the flood pattern was left unchanged over most of the project life. In such cases, there is little difference between the parameter values determined from core data and the values obtained by least-squares fitting of the production response curves. Each waterflood response curve requires only a single set of parameters.

In the curve-fitting exercise described above, it was necessary, conceptually, to associate two different pore volumes with each contaminant concentration vs. time curve in order to match the model to the field data. An analogous situation sometimes occurs in highly stratified petroleum reservoirs, when the system responds as two distinct layers connected only through the wellbores. It is then easy to justify the addition of two response curves to obtain the total production.

In the present case, this justification is not so easy. While the introduction of a mass source into the system by injecting air will certainly change the flow patterns, the treatment of the subsequent behavior as a continuation of the early time performance with a second response added is basically an *ad hoc* assumption and needs to be investigated further. Simulators will be very useful in this investigation, because it is

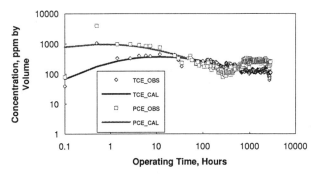

Figure 12.   Final results of fitting field data as the sum of two model curves for each contaminant. The parameter values are given in Table I.

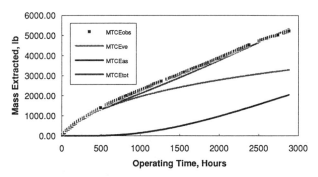

Figure 13.   Comparison of model and field results for cumulative TCE extraction. The lower curve is an estimate for extraction assisted by air stripping, and the next curve above is for extraction alone. The upper smooth curve is the total extraction predicted by the model, which is in close agreement with the data points (shown as squares).

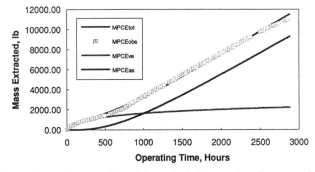

Figure 14.   Comparison of model and field results for cumulative PCE extraction. The curves are similar to those for TCE in Figure 13, except that the curve including the effect of air-stripping crosses the extraction-only curve at about 1000 hours and accounts for most of the PCE by the end of the test.

then possible to specify exactly the distribution of the important physical variables, in contrast to a real system.

However, at present, it is interesting to speculate on the meaning of the results obtained above. One of the difficult problems in analyzing the Savannah River experience is the separation of the effects of the horizontal well geometry from the effects of air injection on the overall VOC recovery performance. Until we can develop a clear rationale for this separation, it is difficult to transfer the technology. Another problem arises in comparing in-situ air stripping with other remediation technologies which have been applied at the site. The pump and treat system has been in operation for more than seven years, but it encompasses a much larger area. For this reason, it would be expected to perform more poorly than a small-scale system located near a "hot spot" of contamination. The vertical-well vacuum extraction test conducted before the horizontal system was developed was also in the hot spot area, but the test was much too short to allow a good comparison to be made with in-situ air stripping.

If the model results can be taken seriously, the short- and long-term performance with and without air injection can be determined directly from the two parts of the model calculation. The model result for total VOC extracted during the entire test is 16834 lb., of which 5501 lb. is attributed to vacuum extraction alone and 11333 lb. is attributed to the increase resulting from air injection. This represents a 200% increase in the amount of contaminant removed for the same operating period.

It is worth noting that at least the qualitative effects of heterogeneity are captured by this simple model, in that very long time periods will be required to achieve regulatory compliance through continued extraction and air injection. Table II shows this dramatically, with the predicted percentage of mass recovered (according to the model) increasing very slowly with continued operating time. Concentrations of contaminants in the extracted air are predicted to *decrease* very slowly, with more than 100 years required to reduce TCE to less than 10 parts per billion in the vapor.

Table II. Extrapolated remediation performance based on the heterogeneity parameter model

Years of Operation	Per Cent TCE Removed	Concentration (ppb)	Per Cent PCE Removed	Concentration (ppb)
1	59.4	56610	51.6	142296
10	90.7	988	97.6	1927
100	98.1	30	>99.9	2
150	>99.0	9	>99.9	<1

Finally, it should be pointed out that this approach, if validated by future work, very neatly separates the effects of heterogeneity from the question of uncertainty. The long period required for remediation is a consequence of heterogeneity, not uncertainty. The only uncertainty is in how serious the problem will be.

Acknowledgments

The author gratefully acknowledges the support and interest of Jesse L. Yow, wishes to thank Dawn Kaback, Brian Looney, and others at the Savannah River Site for many helpful discussions and for supplying data on the Savannah River Integrated Demonstration Project. I am particularly indebted to Terry Walton and John Steele of Westinghouse Savannah River Corporation for their patience in adjusting to numerous delays.

Financial support from the U.S. Department of Energy Office of Technology Development is also gratefully acknowledged. This work was performed at Lawrence Livermore National Laboratory under the auspices of the U.S. Department of Energy contract number W-7405-ENG-48.

Literature Cited

1.   Chesnut, D. A.; Cox, D.O.; and Lasaki, G. *Proceedings, Pan American Congress of Petroleum Engineering*; Petroleos Mexicanos: Mexico City, **1979**; "A Practical Method for Waterflood Performance Prediction and Evaluation."

2.   Chesnut, D.A. *Proceedings, Third International High-Level Radioactive Waste Management Conference*; American Nuclear Society, Inc.: Las Vegas, NV, **1992**; pp 1026-1039

3.   Falta, Ronald W.; Pruess, Karsten; and Chesnut, Dwayne A. *Proceedings, AIChE Summer National Meeting, Advances in Soil Venting, Session 54*; AIChE: Minneapolis, MN, **1992**; "Analytical Modeling of Advective Contaminant Transport During Soil Vapor Extraction."

4.   Kaback, D.S.; Looney, B.B.; Corey, J.C.; and Wright, L.M. III. *Well Completion Report on Installation of Horizontal Wells for In Situ Remediation Tests*, WSRC-RP-89-784; Westinghouse Savannah River Company: Savannah River Site, Aiken, SC; **1989.**

5.   Kaback, D.S.; Looney, B.B.; Corey, J.C.; Wright, L.M. III; and Steele, J.L. *Horizontal Wells for In-Situ Remediation of Groundwater and Soils*; Westinghouse Savannah River Company: Savannah River Site, Aiken, SC, **1989.**

6.   Buscheck, Thomas A.; and Nitao, John J. *Feasibility of In Situ Stripping of Volatile Organic Compounds at the Savannah River Site: Preliminary Modeling of a Pair of Horizontal Wells*; Lawrence Livermore National Laboratory: Livermore, CA., **1992.**

7.   Looney, B.B.; Hazen, T.C.; Kaback, D.S.; and Eddy, C.A. *Full Scale Field Test of the In-Situ Air Stripping Process at the Savannah River Integrated Demonstration Test Site (U)*; WSRC-RD-91-22; Westinghouse Savannah River Company: Savannah River Site, Aiken, SC **1991.**

RECEIVED October 14, 1993

Chapter 3

Electrokinetic Remediation of Unsaturated Soils

E. R. Lindgren[1], E. D. Mattson[2], and M. W. Kozak[1]

[1]Sandia National Laboratories, P.O. Box 5800, Albuquerque, NM 87185
[1]Sat-Unsat, Inc., 12004 Del Rey NE, Albuquerque, NM 87122

An electrokinetic process for *in situ* remediation of anionic contaminants from unsaturated soil is under development at Sandia National Laboratories in Albuquerque, New Mexico. The electromigration rate of anionic food dye ions through unsaturated sand under constant current conditions was measured at various moisture contents ranging from 4 wt% to 27 wt%. The results indicate a maximum migration rate at moisture contents between 14 wt% and 18 wt% water. The drop in the electromigration rate at lower moisture contents is explained by rapid increases in pore tortuosity. The decrease of the migration rate at higher moisture contents is explained by the expected decrease in the pore water current density.

In an experiment using chromate ions, the chromate moved faster than dye ions in an analogous experiment. This was expected because chromate is a smaller molecule. When the chromate reached the graphite anode, it apparently migrated into the slightly porous graphite.

Heavy-metal contamination of soil and groundwater is a widespread problem in the DOE weapons complex, and for the nation as a whole. Large spills and leaks can contaminate both the soil above the water table as well as the aquifer itself. Smaller spills or spills in arid regions with thick vadose zones can result in a contaminant plume that totally resides in unsaturated soil. In both cases there will exist a contamination problem in the vadose zone in need of remediation. Excavation of such sites may not be cost effective or politically acceptable. Electrokinetic remediation is one possible technique for *in situ* removal of such contaminants from unsaturated soils.

In electrokinetic remediation, electrodes are implanted in the soil, and a direct current is imposed between the electrodes. The application of direct current

0097–6156/94/0554–0033$08.00/0

leads to a number of effects: ionic species and charged particles in the soil water will migrate to the oppositely charged electrode (electromigration and electrophoresis), and concomitant with this migration, a bulk flow of water is induced, usually toward the cathode (electroosmosis) (1,2). These phenomena are illustrated in Figure 1. The combination of these phenomena leads to a movement of contaminants toward the electrodes. The direction of contaminant movement will be determined by a number of factors, among which are type and concentration of contaminant, soil type and structure, interfacial chemistry of the soil-water system, and the current density in the soil pore water. Contaminants arriving at the electrodes may potentially be removed from the soil by one of several methods, such as electroplating or adsorption onto the electrode, precipitation or co-precipitation at the electrode, pumping of water near the electrode, or complexing with ion-exchange resins.

Electrokinetic phenomena have been studied for a long time (3). Electrokinetic techniques have been used extensively for the stabilization of soft soils (4,5) and other dewatering operations (6-9) (Lockhart, N. C., *unpublished Proceedings of Workshop on Electrokinetic Treatment and Its Application In Environmental Problems*, University of Washington, August 4-5, 1986.). The notion that electrokinetic phenomena may be applicable to hazardous waste remediation appears to stem from the work of Segall, *et al.* (10), who found that electrokinetically dewatering dredging sludges led to extracted water rich in heavy metals. Ironically, Segall, *et al.* considered this to be an inhibition to the use of electrokinetics in the field, since it produced a toxic liquid waste that required further treatment.

By the mid-1980's, numerous researchers, apparently simultaneously, realized that separations of heavy metals from soils posed a potential contamination solution rather than a potential contamination problem. Mitchell (Mitchell, J., *unpublished Proceedings of Workshop on Electrokinetic Treatment and Its Application In Environmental Problems*, University of Washington, August 4-5, 1986.) and Renaud and Probstein (11) described the possibility for removing contaminants by electroosmosis from fine-grained saturated soils. Shapiro *et al.* (12) have described removing small organic contaminants from columns of saturated clays in the laboratory. In similar bench-scale experiments, Acar, *et al.* (13) and Hamed *et al.* (14) describe the removal of heavy metals from clays. In each of these experiments the concept has been to convect contaminants with water using electroosmosis as the separation mechanism. Runnells and Larson (15) demonstrated the removal of cationic copper from sands in a bench-scale experiment, without considering whether electroosmosis or electromigration caused the removal. So far there have been few field demonstrations of electrokinetic remediation. Lageman (16) has described successful field-scale attempts to remove heavy metals from saturated soils in The Netherlands. A field-scale trial funded by EPA for the removal of chromium contamination from soils has met with partial success, but the success was limited by inadequate site characterization (Banerjee, S., personal communication, 1988).

All of the above studies were conducted using saturated soils. Little study has been conducted toward unsaturated soils. In one recent study (17), the existence of electroosmotic flow in partially saturated clay was demonstrated experimentally. In the Soviet Union (*18*) and in India (*19*), electromigration has been used for remote exploration of some metals in unsaturated soil.

An experimental program is underway at Sandia National Laboratories (SNL) to determine the feasibility of using electrokinetic processes to remediate a chromate plume located in a thick vadose zone beneath the SNL Chemical Waste Landfill (CWL). Past disposal practices have resulted in a chromate plume which was detected at a depth of 75 feet in 1987 as illustrated in Figure 2. Most of the plume is less than 200 parts per million by weight (ppm_w) with a small lens of higher concentration just below the old pit boundary. The plume is entirely in the vadose zone and the aquifer is located 480 feet below the surface. Moisture contents of the soil typically range from 3 to 10 wt%.

In our previous studies, the electromigration of chromate ions and anionic dye ions through unsaturated soil has been demonstrated (*20,21*) This paper reports on a series of experiments that were conducted to study the effect of moisture content on the electromigration rate of anionic contaminants in unsaturated soil, determine the limiting moisture content for which electromigration occurs in a particular soil and compare the migration behavior of chromate ions with a surrogate dye ion.

Experimental

Both sodium dichromate and an anionic food dye were used as contaminants in the experiments presented in this paper. For most of the experiments, pure FD&C Red No. 40 was used as a surrogate for chromium ions. The structures of FD&C Red No. 40 is shown in Figure 3. The dissociation behavior of this molecule can be estimated by assuming one of the sulfate groups behaves similarly to naphthalenesulfonic acid with a $pK_1 = 0.57$ and the other sulfate group behaves like benzosulfonic acid with a $pK_2 = 0.70$ (*22*). This infers that the dye is completely dissociated even at low pH. Chromic acid has a $pK_1 = 0.74$ and a $pK_2 = 6.49$.(*23*) Thus, in neutral to slightly basic pore water (typical of New Mexico), the dye and chromate should be similarly dissociated. However, it is recognized that chromate electromigration behavior will likely be more effected by pH than will the dye analog.

The experimental test cell was constructed of plastic with internal dimensions of 15.2 cm high, 25.4 cm wide and 1.9 cm thick as shown schematically in Figure 4. The electrodes used were graphite rods 1.9 cm square and 15.2 cm high. The electrodes were located at each end of the test cell and the geometry of the electrodes and the cell was such that the soil only contacted one face of each electrode.

The soil used was obtained from near the CWL and wet sieved to retain the 150 to 300 μm diameter fraction. The washed soil was soaked in successive aliquots of synthetic groundwater (0.005 M $CaCl_2$ in distilled water) until the conductivity of the supernatant stabilized. It was then assumed that any adsorption reactions between the soil and the $CaCl_2$ solution had equilibrated. The soil was then desaturated on a pressure plate operated at about 6 bar which reduced the soil moisture content to about 4 wt%. This method of drying was used to prevent the accumulation of $CaCl_2$ that would occur if the soil was oven dried. The moisture

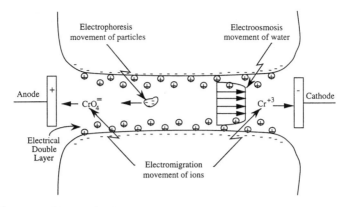

Figure 1.   Electrokinetic phenomena pertinent to in situ remediation.

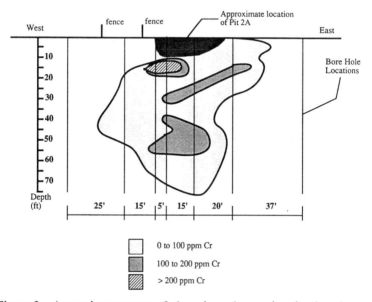

Figure 2.   Approximate extent of chromium plume migration based on a 1987 borehole study.

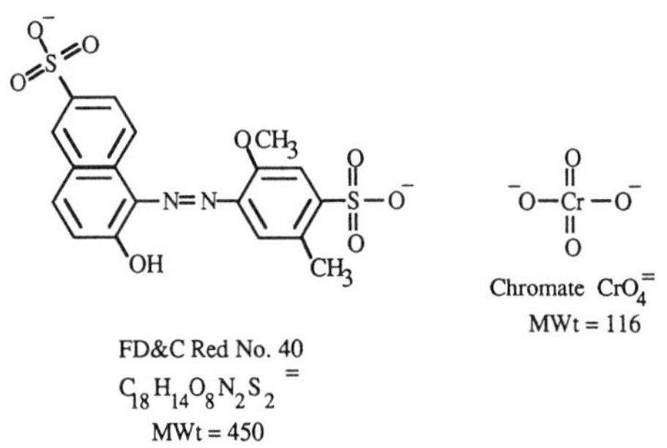

Figure 3.   Chemical structures of FD&C Red Dye No. 40 and chromate ion.

content of the desaturated soil was determined by microwave drying a representative sample. Finally, the appropriate amount of synthetic groundwater was then added back to the soil to produce the desired moisture content for the run.

The soil was packed in the cell while in a horizontal orientation using a 12 ton hydraulic press. After packing the cell, a 0.6 cm wide channel was cut along the center line of the cell, parallel to the electrodes. This channel was filled with soil contaminated with FD&C No.40 or sodium dichromate. The moisture content of the contaminated soil was prepared to match the rest of the soil.

For the series of experiments described in this paper, the molar concentration of contaminant (either dye or chromium) in the pore water was held constant at 0.02 M for the various moisture contents tested. For electrokinetic processes, the pore water concentration is an important parameter. The soil was then treated with a constant current of 10 mA for up to 24 hours with the cell laying flat in a horizontal position to minimize the redistribution of water due to gravity. For the runs using dye, photographs of the dye location were taken at one hour intervals. The advantage of these dye visualization experiments is that the dye location can be monitored *in situ* by photography and no chemical analyses are required. Following each run the cell was dissected and analyzed for moisture content. The average moisture content across the cell was used to define the moisture content of the run.

For the chromate run, after 2, 5 and 9 hours of operation, a 3 cm high horizontal strip (20%) of the soil was removed (refer to Figure 4) and sectioned into 10 to 12 samples for analysis. The remaining 6 cm of soil was sectioned and analyzed after 22 hours of operation. The width of each sample between the center line and the anode where the chromium was expected to be found was 1.5 cm. After removal of each soil strip, the current was reduced proportionally to keep the current density constant throughout the run. Soil samples were prepared for chromium analysis by adding deionized water to each of the soil samples, stirring, and allowing the mixture to settle for one hour. The supernatant water from each sample was drawn into a syringe and then filtered through a 0.45 μm PTFE syringe filter. The filtered water was then analyzed for chromium with an emission spectrometer. The detection limit for the technique used was approximately 0.05 ppm$_w$ (in solution).

Results and Discussion

Contaminant Position vs Time. A typical dye run is depicted in Figure 5 showing the position of the dye initially, and after 8 and 13 hours of operation at 10.2 mA. In these photos, the anode is on the left and the cathode is on the right. In most of the dye runs (especially at lower moisture contents), the trailing edge of the dye was much more distinct than the leading edge. Because of the distinctness of the trailing edge, this boundary was used to define the migration rate. The relative position of the dye trailing edge as a function of time is plotted in Figure 6 for various moisture contents. The run with chromium is also depicted. A common characteristic of all the curves in this figure is the slow initial rate (as indicated by the slope of the curve) followed by a faster rate which remains constant until the end of the run.

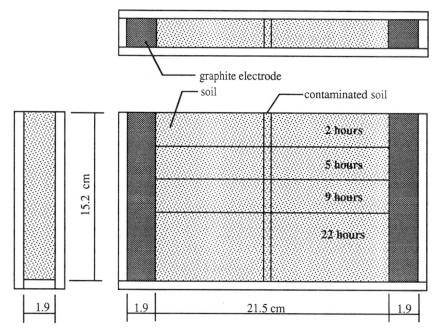

Figure 4. Schematic of test cell set up.

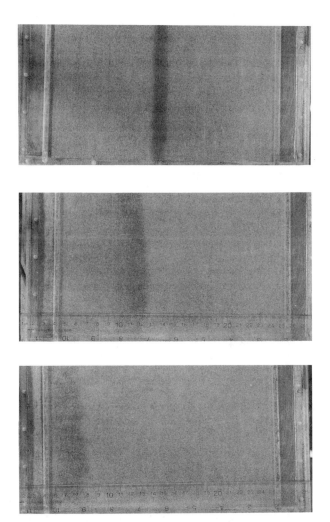

Figure 5. Dye migration through unsaturated soil with 14 wt% moisture.

The "terminal" migration rate for each run was taken to be the slope of the latter portion of each curve as determined by a least square fit. Figure 7 shows a plot of the terminal migration rate as a function of moisture content for all the runs in this study. The results in Figure 7 appear to indicate for this soil an optimum moisture content between 14 and 18 wt% exists for electromigration. Above this moisture content, the rate drops off slightly and below this moisture content, the rate drops off sharply.

It is possible to rationalize the contaminant migration behavior by considering the functional dependence of the electromigration velocity on the ionic concentrations and volumetric moisture content in our unsaturated soil experiments. In these constant current experiments, the planar electrodes produce a one-dimensional voltage gradient which varies only between the electrodes (along the z axis). Following the notation of Probstein (23) consider an ion in dilute solution within a single pore where the x coordinate follows the tortuous pore path:

$$v_{xi} = -v_i F z_i \, dV/dx \tag{1}$$

where:

v_{xi} is the electromigration velocity (dx/dt) of species i along the pore path, ms^{-1}
v_i is the absolute mobility of species i, mole m N^{-1} s^{-1}
F is Faraday's constant, C $mole^{-1}$
z_i is the charge number of species i
dV/dx is the local voltage gradient along the pore path, V m^{-1} or N C^{-1}

Shapiro (24) extended this relationship to porous media using a capillary tube bundle model. This model incorporated pore tortuosity, τ, which accounts for the actual pore path length being longer than through a straight pore which is parallel to the voltage gradient.

$$v_{zi} = -(v_i F z_i / \tau^2) dV/dz \tag{2}$$

where:

v_{zi} is the observed electromigration velocity (dz/dt) of species i along the z axis, m s^{-1}
τ is the pore tortuosity and is defined as dx/dz (24)
dV/dz is the local voltage gradient along the z axis which runs perpendicular to both electrodes, V m^{-1} or N C^{-1}

Assuming no surface conduction and neglecting diffusion, the voltage gradient can be expressed as:

$$dV/dx = \rho_w i_w \tag{3a}$$

or

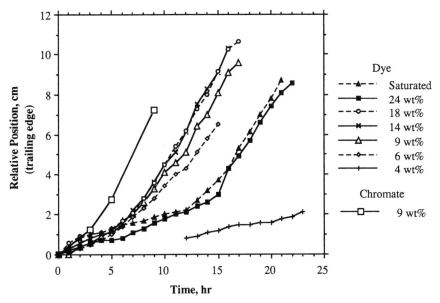

Figure 6. Location of the trailing edge of contaminant versus time for various moisture contents.

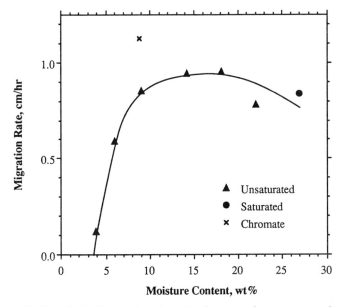

Figure 7. Terminal electromigration velocity vs moisture content for the dye and chromate ions.

$$dV/dz = \tau \, \rho_w \, i_w \tag{3b}$$

where:

ρ_w is the pore water resistivity, ohm m, or N s m^2 C^{-2}
i_w is the pore water current density, A m^{-2}

Assuming all of the current flows through the pore water, the pore water current density can be defined as a function of the moisture content:

$$i_w = I/(A\theta), \tag{4}$$

where:

I = Total current (held constant in these experiments), Amps
A = cross sectional area of test cell perpendicular to current flow, m^2
θ = volumetric moisture content, m^3/m^3

Equation 4 assumes that the water area per unit soil area is equal to the water volume per unit soil volume. This has been shown to be true for voids when pores are randomly oriented (25).

Combining Equations 2, 3b, and 4 and rearranging yields

$$v_{zi} = -(I \; F \; z_i v_i/A)(\rho_w/\tau \, \theta) \tag{5}$$

For a given ion, the first set of terms is constant and Equation 5 shows that the observed electromigration rate is proportional to the pore water resistivity and inversely proportional to the moisture content and tortuosity. The tortuosity of the water filled pore space in unsaturated soil is a function of inverse moisture content ($\tau = f\{1/\theta\}$) (26) and is expected to be highly non-linear.

Thus, there are at least two competing factors contributing to the moisture content dependence of the electromigration velocity. One is the pore water current density, which for constant current conditions increases as the moisture content decreases, and results in a proportionally larger driving force for electromigration. The other is the tortuosity, which also increases as the moisture content decreases, but this decreases the observed electromigration rate in a highly non-linear fashion (Equation 5). The results shown in Figure 7, assuming the concentration changes of the dye are similar in all the runs, suggest that near saturation, the pore water current density effect dominates the tortuosity effect and the electromigration rate increases as the moisture content decreases. However, at lower moisture contents, the non-linear tortuosity effect appears to dominate and the electromigration rate decreases sharply as the moisture content decreases further.

Note that the extrapolated zero electromigration rate occurs at a moisture content of 3.5 wt%. This appears to correlate with the residual moisture (or lowest moisture) portion of the moisture characteristics curve which is shown in Figure

8 for this soil. The residual moisture is also called immobile water (*26*) and is attributed to a loss of pore water connectivity. This loss in pore water connectivity should result in a break of the electrical path through the pore water. Also, it is reasonable to expect electromigration to cease when connectivity of the pore water is completely lost. The point at which connectivity is lost is primarily a function of the soils pore size distribution, and mineralogy. Thus, it may be possible to predict the minimum moisture conditions at which electrokinetic remediation will operate by conducting this simple hydraulic characterization test of the soil.

The pore water resistivity should not in general be a function of the moisture content of the soil, but it is a function of the pore water chemistry which may change with time and position. The more ions in the pore water, the lower the pore water resistivity. The pore water resistivity can be calculated from the concentration and mobility of each ion in solution (*23,27*):

$$\rho_w = 1/(F^2 \sum z_i^2 \, c_i \, v_i) \tag{6}$$

where:

c_i = the molar concentration of species i, mole m^{-3}

Equations 5 and 6 suggest that, in a simplified sense, the pore water resistivity, and hence the electromigration velocity of an ion is dependent on the total ionic concentration and not necessarily on the concentration of the contaminant ion itself. If the concentration of the contaminant ion is significant in comparison to the total concentration of ions in the pore water, then, the contaminant concentration will affect the electromigration rate. However, as the concentration of a contaminant drops such that it is no longer significant in comparison to the total concentration of ions in the pore water, then, further drops in contaminant concentration will not effect the rate of electromigration. In the experiments presented in this paper, the initial contaminant concentration was significantly greater than the concentration of CaCl$_2$ in the pore water solution. As the experiment progressed, the contaminant concentration perhaps dropped below the concentration of CaCl$_2$ (the contaminant concentration was only measured in the chromate run where it dropped to half the CaCl$_2$ concentration after 3 hours). This may explain why the results in Figure 6 show all the runs reach a constant, maximum rate in the latter part of the experiments.

Chromate Electromigration. The results of the run using chromate ions as the contaminant are presented in Figure 9. In this figure, the concentration of chromium in the soil is plotted as a function of position in the test cell for the initial state and at each of the times the cell was sampled. The figure shows that the chromate migrated towards the anode with time. Similar to the results of the dye experiments, the trailing edge of the chromate migrated faster between the 5 and 9 hour samples than it did in the first 5 hours of the experiment. Also note that the concentration of the chromate decreased almost an order of magnitude and the area of soil containing chromate widened while it migrated from the narrow region initially spiked toward the anode. When the chromate reached the anode, the concentration increased as the area of contamination again narrowed.

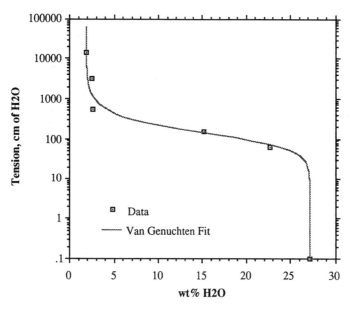

Figure 8. Moisture characteristics curve for the CWL soil used in this study.

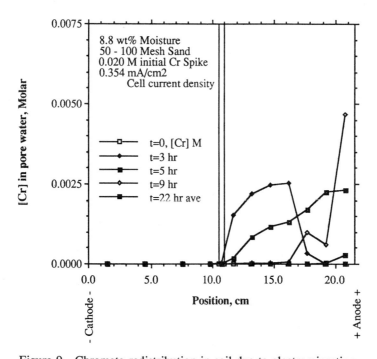

Figure 9. Chromate redistribution in soil due to electromigration.

The chromium mass balance for the first two samples was good with 79% recovered in the sample take at 3 hours and 84% recovered in the sample taken at 5 hours. Note that at the 5 hour sampling, the leading edge of the chromate may not have yet reached the anode because the [Cr] at the leading edge is comparable with that at the leading edge in the previous sampling. At the 9 hour sampling, the leading edge of the chromate had apparently reached the anode because the [Cr] at the anode is twice that of the previous two samples. The Cr mass balance at the 9 hour sampling was only 56% which suggests that when the chromate reaches the anode and the concentration begins to build, the Cr^{+6} begins to change form. By the 22 hour sampling, very little Cr remained detectable by the water extraction technique used. At the anode, the chromate possibly adsorbs onto the graphite surface or is reduced to Cr^{+3} and adsorbs to the soil. Either of these changes would result in no Cr being extracted by the water extraction technique used. Huang and Wu (*28*) have reported high efficiencies in Cr^{+6} removal from dilute aqueous solutions by activated carbon. They found that under acidic conditions, all the Cr^{+6} not adsorbed onto the activated carbon was reduced in solution to Cr^{+3}. The optimal pH was determined to be between 5 and 6 at which nearly all the Cr was adsorbed. While graphite is not the same as activated carbon with respects to structure and surface area, it is probable that the carbon in each react similarly with chromate. The pH of the soil in the chromate run was not measured, but in the analogous dye run after 22 hours of 10.2 mA the soil near the anode exhibited a soil pH of 5.

In an effort to determine where the Cr went, the surface of the graphite electrode was analyzed by a couple of surface analyzing techniques, electron microprobe and secondary ion mass spectrometry (SIMS). An analysis of a cross section of the electrode by electron microprobe indicated chromium had migrated at least 400 μm into the electrode. The chromium concentration ranged from about 100 ppm$_w$ at 375 μm to about 500 ppm$_w$ at 50 μm. These concentrations of chromium in the electrode can account for much of the missing chromium. SIMS detected small amounts of Cr to even greater depths, with more significant amounts of Al, Si, K, Ca, Fe and Mg being detected.

The moisture content of each sample was determined as part of the chromium analysis. Figure 10 shows the moisture content as a function of position for each of the times sampled. Electroosmosis causes water to move away from the anode toward the cathode. An estimate of 0.03 cm/hr for the rate of electroosmosis can be made from a water balance on the samples taken adjacent to the electrodes. It was assumed that all of the moisture content change was due to electroosmotic flow. This assumption implies the rate of hydrolysis of water at the electrodes and the counter unsaturated counter flow are negligible. These assumptions are supported by complete recovery of water for the first 9 hours of the run and 97% recovery of water after 22 hours. It is unlikely that any significant unsaturated hydraulic flow occurred due to the low unsaturated hydraulic conductivity and pressure head gradients. In both the chromate and the dye experiments, the direction of electromigration was opposite the direction of electroosmotic flow. However, there appears to be little electroosmotic effect because the magnitude of the electromigration velocity is many times the magnitude of the electroosmotic flow velocity.

It is possible to estimate the average pore water resistivity using Equation 5 for comparison with the average electromigration rate calculated from the trailing edge location in each sampling. Figure 11 shows a plot of the chromium electromigration rate versus the calculated pore water resistivity. The line is the least square fit to the limited data which passes surprisingly close to the origin. As suggested by Equation 5, the rate of electromigration should be proportional to the pore water resistivity as indicated by this plot. The necessary assumptions are: 1) the moisture content remains constant, 2) the calcium and chloride ion concentrations (which were not measured) remained constant, 3) all the chromium from the sodium dichromate is present as chromate ions, 4) the sodium ions are in the pore water with the chromate (to maintain electroneutrality), 5) the chromate ion concentration is the average of the trailing edge concentration measured in two successive samples and 6) calcium, sodium, chloride and chromate are the only ions in the pore water and their mobility can be estimated from equivalent ionic conductivities at infinite dilution (27). The measured resistivity of the 0.005 M $CaCl_2$ solution used in this study was 8.2 ohm m; the calculated value for infinite dilution is 7.4 ohm m and for 0.005 M is 8.1 ohm m. This suggests that the method used for calculating the pore water resistivity is accurate and assuming infinite dilution does not lead to gross errors.

Thus, for the chromate run, the increase in the rate of electromigration with time evident in Figure 6 is explained by changes in the pore water resistivity due to the drop in the chromate concentration as the plume spread out. The constant, maximum electromigration rate observed in the dye experiments are probably due to similar effects. This would infer that the dye concentration either reached a constant value or dropped to a point where further changes did not affect the pore water resistivity.

Comparison of Cr and Dye Electromigration Rates. A comparison of the electromigration rates of the chromate ions and the dye ions can be made. The chromate position of the trailing edge is plotted on Figure 6 along with all of the dye data. The chromate ions exhibit very similar behavior as the dye. Although the runs with chromate require sampling at a more course scale, it is still apparent that the migration rate of both the dye and the chromate ions are initially slower than the terminal rate. However, the chromate ions reached the maximum rate sooner than the dye so that the overall migration rate of the chromate ions was about 2.2 times faster than the dye at a similar moisture content. The terminal migration rate of the chromate is plotted on Figure 7 along with all of the dye data. The terminal migration velocity of the chromate ions is 1.3 times faster than the terminal dye migration rate at a similar moisture content.

Conclusions

The simple model for electromigration in unsaturated soil developed in this paper predicts dependence on soil moisture content and pore water chemistry which are evident in the experiments conducted. This suggests that the experimental techniques employed are useful in studying fundamental aspects of electromigration of anionic contaminants in unsaturated soil.

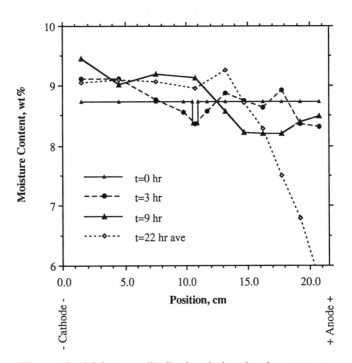

Figure 10. Moisture redistribution during the chromate run.

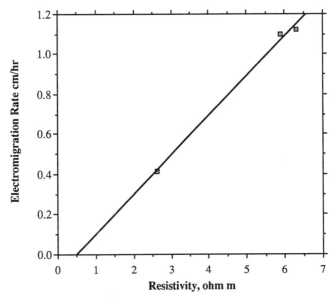

Figure 11. Electromigration rate of chromate ions versus calculated pore water resistivity.

The preliminary results appear to indicate that for the soil used, the optimum moisture content for electromigration is less than saturation due to competing effects of tortuosity and pore water current density. The minimum moisture content for electromigration in the soil was very low and may be estimated from the residual moisture content of the soil.

Chromate and dye ions exhibited similar migration behavior. Like the dye, the initial migration rate of the chromate is slower than the final migration rate. However, the chromate migrated faster than the dye at similar moisture contents. The qualitative similarities between the dye and the chromate ions make the dye a useful surrogate for chromate ions.

As the chromate reaches the graphite anode, it apparently changes valence and either adsorbs onto the graphite or the soil adjacent to the electrode as Cr^{+3}. It may be possible to optimize electrode conditions to promote the adsorption of chromate onto the electrode. This phenomena may be utilized in conjunction with electromigration as a method to selectively remove chromate from unsaturated soils.

Notation

A	cross sectional area of test cell perpendicular to current flow, m2
c_i	the molar concentration of species i, mole m^{-3}
dV/dx	local voltage gradient along the pore path, V m^{-1} or N C^{-1}
dV/dz	local 1-D voltage gradient along the z axis which runs perpendicular to both electrodes, V m^{-1} or N C^{-1}
F	Faraday's constant, C $mole^{-1}$
I	total current (held constant in these experiments), Amps
i_w	pore water current density, A m^{-2}
v_{xi}	electromigration velocity (dx/dt) of species i along the pore path, m s^{-1}
v_{zi}	observed electromigration velocity (dz/dt) of species i perpendicular to the potential gradient, m s^{-1}
z_i	charge number of species i
θ	volumetric moisture content, m^3/m^3
ρ_w	pore water resistivity, ohm m, or N s m^2 C^{-2}
τ	pore tortuosity and is defined as dx/dz (24)
υ_i	absolute mobility of species i, mole m N^{-1} s^{-1}

Acknowledgments

This work was funded by the Office of Technology Development, within the Department of Energy's Office of Environmental Restoration and Waste Management, under the Mixed Waste Landfill Integrated Demonstration and the In Situ Remediation Technology Development Integrated Program.

This work was performed at Sandia National Laboratories, which is operated for the U.S. Department of Energy under Contract No. DE-AC04-76DP00789.

Literature Cited

1. Overbeek, J. T. G., and B. H. Bijsterbosch, in *Electrokinetic Separation Methods*, Righetti, Van Oss, and Van der Hoff, eds., Elsevier, 1979.

2. Hunter, R. L., *Zeta Potential In Colloid Science*, Academic Press, New York, 1981.

3. Reuss, F. F. *Memoires de la Societe Imperiale des Naturalistes de Moscou*, **1809**, *2*, pp. 327.

4. Casagrande, L. *Geotechnique* **1949**, *1*, pp. 1959.

5. Nikolaev, B. V. *Pile Driving By Electroosmosis*, Consultants Bureau, 1962.

6. Sprute, R. H., and Kelsh, D. J., *Bureau of Mines Report of Investigations*, RI8441, U.S. Department of the Interior, 1980.

7. Sprute, R. H., and Kelsh, D. J., *Bureau of Mines Report of Investigations*, RI8666, U.S. Department of the Interior, 1982.

8. Lockhart, N. C., *Colloids and Surfaces*, **1983**, *6*, pp. 239.

9. Sunderland, J. G., *J. Applied Electrochemistry*, **1987**, *17*, pp. 889.

10. Segall, B. A., O'Bannon, C. E., and Matthias, J. A., *J. Geotech Engineering Division ASCE*, **1980**, *106*, No. GT10, pp. 1148-1153.

11. Renaud, P. C., and Probstein, R. F., *Physico-Chemical Hydrodynamics*, **1987**, *9*, pp. 345.

12. Shapiro, A. P., P. C. Renaud, R. F. Probstein, *Physico-Chemical Hydrodynamics*, **1989**, *11*, pp. 785-802.

13. Acar, Y. B., Gale, R.W., Putman, G., Hamed, J., "Electrochemical Processing of Soils: Its Potential use in Environmental Geotechnology and Significance of pH Gradients" *Second International Symposium on Environmental Geotechnology*, Shanghi, China, May 1989.

14. Hamed, J., Acar, Y., Gale, R., *J. of the Geotechnical Engineering Division ASCE*, **1989**.

15. Runnells, D.D., J.L Larson, *Ground Water Monitoring Review*, **1986**, *6*, 3, pp. 85-88.

16. Lageman, R., Pool, W., & Seffinga, G., *Chemistry and Industry*, September 18, **1989**, pp. 585-590.

17. Mitchell, J. K., A. T-C. Yeung, "Electrokinetic Flow Barriers in Compacted Clay", *Transportation Research Board*, 69th Annual Meeting, Washington, D.C., Paper No. 890132, January, 1990.

18. Shmakin, B., *J. Geochemical Exploration*, **1985**, *23*, pp. 27-33.

19. Talapatra, A. K., R. C. Talukdar, P. K. De, *J. of Geochemical Exploration*, **1986**, *25*, pp. 389-396.

20. Lindgren, E.R., M.W. Kozak, E.D. Mattson, "Electrokinetic Remediation of Contaminated Soils", Proceedings of the ER'91 Conference at Pasco, WA, Sept 1991, pp 151.

21. Lindgren, E.R., M.W. Kozak, E.D. Mattson, "Electrokinetic Remediation of Contaminated Soils: An Update", Proceedings of the Waste Management 92 Conference at Tucson, AZ, March 1992, pp. 1309.

22. Weast, R. C., *CRC Handbook of Chemistry and Physics*, 66th Edition, CRC Press, Boca Raton, FL 1985-1986.

23.   Probstein, R.F., *Physicochemical Hydrodynamics*, Butterworth Publishers, Stoneham, MA, 1989.

24.   Shapiro, A., P., "Electroosmotic Purging of Contaminants from Saturated Soils", PhD Thesis, Massachusetts Institute of Technology, 1990.

25.   Schechter, R. S. and Gidley, J. L., "The Change in Pore Size Distribution from Surface Reaction in Porous Media," *AIChE J.*, **1969**, *15*, No. 3, pp. 339.

26.   Hillel, D, *Fundamentals of Soil Physics*, Academic Press, New York, NY, 1980.

27.   Daniels, F., and R. A. Alberty, *Physical Chemistry*, 4th edition, John Wiley & Sons, New York, NY, 1975.

28.   Huang, C. P. and M. H. Wu, "The Removal of Chromium(VI) From Dilute Aqueous Solution by Activated Carbon", *Water Research*, **1977**, *11*, pp. 673-679.

RECEIVED January 14, 1994

Chapter 4

Determination of Bioavailability and Biodegradation Kinetics of Phenol and Alkylphenols in Soil

Henry H. Tabak[1], Chao Gao[2], Lei Lai[2], Xuesheng Yan[2], Steven Pfanstiel[2], In S. Kim[2], and Rakesh Govind[2]

[1]Office of Research and Development, Risk Reduction Engineering Laboratory, U.S. Environmental Protection Agency, Cincinnati, OH 45268
[2]Department of Chemical Engineering, University of Cincinnati, Cincinnati, OH 45221

Knowledge of biodegradation kinetics in soil environments can facilitate decisions on the efficiency of in-situ bioremediation of soils, sediments and aquifers. This paper reports on a study whose main goal is to quantitate the bioavailability and biodegradation kinetics of organics in surface and subsurface soils and develop a predictive model for biodegradation kinetics applicable to soil systems. The adsorption/desorption equilibria and kinetics from soil particles were measured for phenol and alkyl phenols. Studies were conducted on biodegradation of phenol and alkyl phenols using soil slurry reactors. Measurements of oxygen consumption and carbon dioxide generation were made in an electrolytic respirometer and biokinetic parameters were derived from the data using a mathematical model which incorporates the effects of adsorption/desorption from the soil particles and biodegradation in the liquid and soil phases. Protocols for measuring biomass adsorption in soil, quantifying carbon dioxide evolution using shaker flasks and measurement of radiolabelled carbon dioxide evolution in respirometric flasks are also presented in this paper.

Quantification of biodegradation kinetics can provide useful insights into the range of environmental parameters for enhancing bioremediation rates (1,2). There is increased interest in the development of bioremediation technologies for decontaminating aquifers and soils. Biological treatment systems, especially when used in conjunction with other physical/chemical treatment methods, hold considerable promise for safe, economical, on-site treatment of toxic wastes. A variety of biological treatment systems designed to degrade or detoxify environmental contaminants are currently being developed or marketed by commercial vendors.

0097–6156/94/0554–0051$09.26/0

Irreversible binding of the chemical to the soil phase may limit its bioavailability and ultimate degradation. Recently, increased interest has been directed towards obtaining quantitative information on pollutant sorption equilibria in soils, since the physical state of the compound can influence its bioavailability.

The main objectives of this research are to quantitate the bioavailability and biodegradation kinetics of organic chemicals in surface and subsurface soil environments, examine the effects of soil matrices and soil conditioning (drying, aging, compacting), and develop a predictive model for biodegradation kinetics applicable to soil systems.

This paper highlights biodegradation studies on phenol and several alkyl phenols: p-cresol, 2,4-dimethyl phenol, catechol, hydroquinone and resorcinol. The studies incorporate the use of soil microcosms for acclimation of soil microbiota; measurement of respirometric oxygen uptake and carbon dioxide evolution in soil slurry reactors; measurement of carbon dioxide generation rates in shaker flask systems; and measurement of adsorption/desorption equilibria and kinetics. Studies are also conducted with radiolabelled compounds to quantify the kinetics in soil matricies. A mathematical model for soil slurry reactors is used to determine biodegradation kinetics for the selected compounds from the oxygen uptake data.

Background

Studies on Biodegradation Kinetics. Respirometric technologies for evaluating and testing biodegradation of organic compounds have gained prominence as reliable methods for quantitating the fate of these compounds in aqueous environments (3,4,5,6). Linear and non-linear group contribution models were developed for predicting biodegradation kinetics for organics in aqueous systems (5,6) using biokinetic data, obtained respirometrically.

Extension of the respirometric biokinetics determination protocol to predict the effect of soil on biodegradability of organic pollutants seems to be very desirable. By conducting parallel respirometric biodegradation tests in presence and absence of a soil matrix, it should be possible to determine the biodegradability and biodegradation kinetics of these compounds in soil-slurry systems and in undisturbed soil layers. The resulting information could then be used in combination with groundwater and contaminant transport models to assess the potential efficacy of bioremediation at a particular site.

Recently, several research studies reported the use of respirometry for evaluating biodegradation kinetics of organic pollutants in soil. Long-term respirometric BOD analysis was applied to bench-scale studies to continuously monitor bacterial respiration during growth in mixed organic wastes from contaminated water and soil, in order to assess the potential for stimulating biodegradation of these wastes (7). This information was used to make an initial determination regarding the need to further explore bioremediation as a potential remedial action technology using on-site, pilot-scale testing.

A treatability study used electrolytic respirometers and biometers to determine the biodegradation potential of crude oil petroleum-based wastes (drilling mud, tary material and heavy hydrocarbons) as contaminants of a polluted soil site area (8). The treatability data provided biotreatment efficiencies of the petroleum wastes and were used to ascertain the bioremediation clean-up time.

The use of radioisotopes in biodegradation studies can significantly increase the sensitivity of biodegradation measurements and realistic estimates of biodegradation can be obtained with a minimum of analytical interference from an environmental matrix, at concentrations which are often outside the scope of screening methods. Biodegradation products may be quantified using a combination of separation and radiochemical detection techniques. Recently, techniques were developed for the measurement of Monod kinetic coefficients, microbial yield and endogenous decay coefficients, through measurement of initial rates in short duration batch experiments using radiolabeled substrates (9).

The application of radiolabeled substrate techniques has recently been expanded to the determination of biodegradation of organic compounds in soil. Radiolabelled techniques were used to study the persistence of pentachlorophenol (PCP) and mercuric chloride ($HgCl_2$) in soil, their effect on the microbiota and the reversibility of that damage in numerous soil types, at different substrate concentrations (10). Studies were also reported on determination of biomineralization rates of ^{14}C-labeled organic chemicals in aerobic and anaerobic suspended soil (11).

Development of biodegradation models for organic pollutants in soil is difficult due to the existence of several complicating factors, which includes: (1) presence of diffusional barriers in soil macro and micro pores for compound, oxygen or nutrients; (2) effect of chemical sorption to clay and humic constituents; (3) presence of other biodegradable organic matter in soil; (4) changes in microbiota growth due to protozoa parasitizing the biodegrading population; (5) effect of compound solubility in the aqueous phase; (6) formation of biofilms on the soil surface in conjunction with suspended cultures; and (7) existence of unacclimated soil microbiota.

Most biodegradation kinetic models have neglected sorption of the contaminant on soil particles, which has been shown to be important in contaminant transport. Kinetically, sorption is a two-phase process, with an initial fast stage (< 1 hour) followed by a slower long phase (days), controlled by diffusion to internal adsorption sites (12).

Soil consists of various size pores, with about 50% of the total pore volume consisting of pores with radii < 1 μm. Most soil bacteria range in size from 0.5 to 0.8 μm, and hence a significant portion of the soil may be inaccessible to most bacteria.

The role of soil aggregates and effect of their characteristics on bioremediation in soil have been analyzed in the contaminated aggregates bioremediation (CAB) model (13). Sensitivity analysis of the CAB model has shown the effects of aggregate radius, partition coefficient, and initial contaminant concentration on the time and mechanism of remediation. Diffusion rates of substrate and oxygen and biodegradation kinetics were found to be controlling mechanisms for remediation in the aggregates.

The kinetics of phenolic compounds removal in soil were studied by Namkoong et al. (14) and zero order and first order kinetic models were evaluated to determine their adequacy to describe the removal of seventeen phenolic compounds in soil. Both models were shown to describe the removal of these phenols from soil. A kinetic model which incorporates microbial growth may be desirable to describe the removal of certain compounds, when removal is associated with growth. Future studies should thus determine the importance of microbial growth in the removal kinetics of hazardous waste constituents in soil systems.

The success of bioremediation processes lies in degrading the organic contaminants and in reducing both the toxicity and migration potential of the hazardous constituents in soil. Laboratory studies were conducted using phenolic compounds (15) to characterize overall chemical degradation and toxicity reduction in a contaminated soil. Results indicated that first-order kinetics described the loss of phenolic compounds satisfactorily and loss of contaminant in the water soluble fraction was faster than loss of the same chemicals in soil. Furthermore, toxicity of the water soluble fraction decreased with concentration and no enhanced mobilization of the applied chemicals was observed during the degradation process.

Alexander and Scow (16) and Scow et al. (17) reviewed biodegradation kinetics in soil and discussed the effects of diffusion and adsorption. They concluded that current models are based on studies of single microbial population or single enzymes and considering the complexity of biodegradation processes in soil, it is unlikely that a single model or equation would be applicable for the degradation of all organic substrates in all types of soil environments.

Abiotic Studies on Adsorption/Desorption Equilibria and Kinetics. It is essential to determine the adsorption/desorption equilibria and kinetics of the organic pollutant compounds in soil matricies while evaluating their biodegradability, in order to quantitate the compound's bioavailability in the soil-water environment and to close the mass balance of all reactions that characterize the environmental fate of these organics in soil.

Studies have been recently reported in the literature on the influence of adsorption and desorption of organic compounds on their biodegradation, toxicity and transport in soil (18,19,20). Bioavailability of soil-sorbed contaminants may have a bearing on the effectiveness of microbial degradation as well as on the assessment of toxicological risks. Studies of Weisenfels et al.(18) to determine the soil characteristics that prevent polycyclic aromatic hydrocarbons (PAH) degradation have shown that migration of PAHs into soil organic matter representing less accessible sites within the soil matrix makes them non-bioavailable and thus non-biodegradable. By eluting soil with water, no biotoxicity, assayed as inhibition of bioluminescence, was detected in the aqueous phase. The data suggest that sorption of organic pollutants into soil organic matter significantly effects biodegradability as well as biotoxicity.

Batch soil microcosms were used by Robinson et al. (19) to evaluate the sorption and bioavailability of toluene in an organic soil containing acclimated bacteria. Most toluene sorption occurs rapidly but a small fraction sorbs at a much slower rate and a true equilibrium may take a sinificant amount of time to achieve. Measurement of both sorbed and solution phase toluene concentrations indicate that acclimated bacteria quickly utilize toluene under aerobic conditions. Desorption of most toluene from soil is rapid, thereby becoming available to acclimated bacteria in the aqueous phase. However, a small quantity of toluene desorbs very slowly and becomes available for biodegradation at a rate limited by desorption. Studies of Weber et al. (20) indicate that the behavior, transport and ultimate fate of contaminants in subsurface environment may be affected significantly by their participation in sorption reactions.

Gordon and Millers (21) reported on adsorption mediated decrease in the biodegradation rate of organic compounds showing that adsorbed organic material is less available to bacterial degradation than the same material in a dissolved state and that the decreased availability is a function of the degree of adsorption. Removal of the substrate from the surface of the soil is the rate limiting step.

Loosdrecht et al. (22) reported on the influence of interfaces on microbial activity (growth rate, yield, substrate conversion and efficiency) and concluded that the presence of surfaces may positively or negatively or not at all affect microbial substrate utilization rates and growth yields. The results often depend on the nature of the organism, the kind and concentration of substrate, and the nature of solid surface. In interpreting the effect of surfaces on bioconversion processes, all possible physical or chemical interactions (diffusion, adsorption, desorption, ion-exchange reactions, conformation changes, etc.) of a given compound and its possible metabolites with a given surface have to be considered before general conclusions can be drawn. Their review implies that there is no conclusive evidence that adhesion directly influences bacterial metabolism. All differences between adhered and free cells, depending on conditions, could all be attributed to an indirect mechanism by which the surroundings of the cells, rather than the cells, are modified due to presence of surfaces.

The effect of simultaneous sorption and aerobic biodegradation in the transport of several dissolved alkyl benzenes in an aquifier material was investigated by Angley et al (23). Evaluation was made on the performance of a coupled-process model. Predictions obtained with the coupled-transport model, wherein sorption was assumed to be rate-limited matched the breakthrough curves better than the predictions obtained with a model wherein sorption was assumed to be instantaneous. Accordingly, the assumptions upon which the use of the model is based, e.g., biodegradation occurs only in solution and without any significant acclimation period, and can be simulated with a first order equation, appear to be realistic.

Studies to identify the process(es) responsible for non-equilibrium sorption of hydrophobic organic compounds (HOCs) by natural sorbents were reported by Brusseau et al. (24). The results of experiments performed with natural sorbents were compared to rate data obtained from systems wherein rate-limited sorption was caused by specific sorbete-sorbent interactions. The comparison showed that chemical non-equilibrium associated with specific sorbate-sorbent interactions does not significantly contribute to the rate-limited sorptions of HOCs by natural sorbents. Transport related non-equilibrium was also shown not to be a factor. Based on the interpretation of data in terms of two, sorption-related, diffusive mass transfer conceptual models, there is strong evidence that intraorganic matter diffusion is responsible for non-equilibrium sorption exhibited by the system investigated.

Methodology

Studies on Soil Microcosm Reactors. The methodology for the establishment of the bench-scale microcosm reactors, including the reactor configuration and supportive equipment description, undisturbed soil cube sampling for inclusion in the microcosms, procedure for contamination of soil bed with homologous series of organic compounds, description of the application of nutrients and the operation of the microcosm reactors (CO_2 generation analysis and chemical analysis of soil samples and reactor leachates for the parent compound and metabolites) has been fully described elsewhere (25,26). A schematic of soil microcosm reactor is shown in Figure 1.

Each microcosm reactor represents a controlled site, which eventually selects out the acclimated indigenous microbial population in the soil for the contaminating organics. Samples of soil are then taken from the microcosm reactors and used as source of acclimated microbial inoculum for measuring: (1) oxygen uptake respirometrically; (2) carbon dioxide generation kinetics in shaker flask reactors; and (3) for studies with other soil reactor systems. The microcosm reactor units are also being used directly to evaluate the biodegradability of the pollutant organics and to measure their average biodegradation rate in this intact, undisturbed soil bed.

Control soil microcosm reactors incorporated in the study are: uncontaminated soil, synthetic soil, and soil contaminated with either a solvent or an emulsifier used to dissolve or emulsify each group of selected compounds.

Uncontaminated forest soil was selected for all studies presented in this paper. Soil characteristics, measured using standard methods (27), are summarized in Table I.

Respirometric Studies with Soil Slurries. Studies were conducted with soil slurry reactors, wherein the oxygen uptake was monitored respirometrically. The extent of biodegradation and the Monod kinetic parameters for variety of the organic pollutant compounds by soil microbiota were determined from oxygen uptake data. Various concentrations of soil (2, 5, 10%) and compound (50, 100, and 150 mg/L) were mixed with a synthetic medium consisting of inorganic salts, trace elements and either a vitamin solution or solution of yeast extract and stirred in the respirometric reactor flasks. The flasks were connected to the oxygen generation flask and pressure indicator cells of a 12 unit Voith Sapromat B-12, electrolytic respirometer, and the oxygen uptake (consumption) data were generated as oxygen uptake velocity curves.

A detailed description of the Voith Sapromat B-12 electrolytic respirometer (Voith Inc., Heidenheim, Germany) has been presented elsewhere (3).

Experimental Approach. The nutrient solution used in the respirometer was an OECD synthetic medium (28,29) containing the following constituents in deionized distilled water: (1) mineral salts solution containing KH_2PO_4, K_2HPO_4, $Na_2HPO_4.2H_2O$, NH_4Cl, $MgSO_4.7H_2O$, $CaCl_2$ and $FeCl_3.6H_2O$; (2) trace salts solution containing $MnSO_4.H_2O$, H_3BO_3, $ZnSO_4.7H_2O$, and $(NH_4)_6Mo_7O_{24}$; (3) vitamin solution containing biotin, nicotinic acid, thiamine, p-aminobenzoic acid, pantothenic acid, cyanocobalamine a d folic acid; and (4) yeast extract solution as a substitute for vitamin solution. T soil served as a source of inoculum. The concentration of forest soil in the reactor flask varied from 2 to 10 % by weight, using dry weight of soil as the basis. The total volume of the slurry in the flask was 250 ml.

A more comprehensive description of the procedural steps for the respirometric tests is presented elsewhere (3). The test and control compound concentrations in the media were 100 mg/L. Aniline was used as the biodegradable reference compound at a concentration of 100 mg/l. The typical experimental system consisted of duplicate flasks for the reference substance, aniline, and the test compounds, a single flask for the physical/chemical test (compound control), and a single flask for toxicity control. The contents of the reaction vessels were preliminarily stirred for an hour to ensure endogenous respiration state at the initiation of oxygen uptake measurements. Then the test compounds and aniline were added to it. The reaction vessels were then incubated at 25°C in the dark (enclosed in the temperature controlled waterbath) and stirred continuously throughout the run. The microbiota of the soil samples used as an inoculum were not pre-acclimated to the substrates. The incubation period of the experimental run was 28 to 50 days.

Kinetic Analysis of Soil Slurry Oxygen Uptake Data. The oxygen uptake data were analyzed by computer simulation techniques and curve fitting methods, using the Monod equation combined with a nonlinear adsorption isotherm, to determine the soil

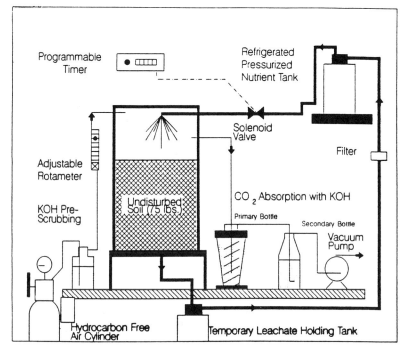

Figure 1.　Schematic of soil microcosm reactor system.

Table I.　Characteristics of soil used in this study

Total Carbon		0.415%
Carbonate		negligible
Organic Carbon		0.415%
Organic Matter		0.80%
Particle size distribution		
< 2	μm	15.65% by weight
2-10	μm	5.00%
10-20	μm	7.75%
20-44	μm	7.90%
44-75	μm	4.55%
75-150	μm	6.50%
150-300	μm	14.30%
300-600	μm	22.10%
600-1180	μm	16.25%

adsorption parameter, and the Monod equation biokinetic parameters. The experimental data were analyzed using a mathematical model (equations given below) which include the effect of chemical adsorption / desorption on the soil particles and the adsorption / desorption of the microorganisms from the soil.

$$X_s = K_b X_w \tag{1}$$

$$S_s = K_d S_w^{1/n} \tag{2}$$

$$O_2 = (S_{to} - S_t) - (X_t - X_{to}) - (S_p - S_{po}) \tag{3}$$

$$dX_w/dt = (\mu_w X_w S_w)/(K_w + S_w) - (b K_w X_t)/(K_w + S_w) \tag{4}$$

$$dS_w/dt = -(1/Y)(\mu_w X_w S_w)/(K_w + S_w) \tag{5}$$

$$dS_p/dt = -Y_p(dS_t/dt) \tag{6}$$

where subscripts s, w and p represent the soil, water and degradation products respectively. S is the concentration of compound, X is the concentration of biomass, and S_p is the conecentration of the degradation products. μ_w, K_w, Y and Y_p are the Monod equation maximum specific growth rate parameter, Michaelis constant, biomass yield and product yield coefficient, respectively. b is the biomass decay coefficient, K_d is the soil adsorption isotherm parameter, (1/n) is the soil adsorption intensity coefficient, and Kb is the biomass adsorption parameter. O_2 is the cumulative oxygen uptake.

The experimental values of the oxygen uptake were matched with the theoretically calculated values and the best fit for the parameters was obtained using an adaptive random search method.

Measurement of Carbon Dioxide Evolution Rates. Carbon dioxide generation rates were measured in shaker-flask soil slurry systems and in the electrolytic respirometry soil-slurry reactors in order to assess the rate and extent of biodegradation/mineralization. The CO_2 generation rate measurement serves as an additional tool for quantitating the biofate of these organics in addition to the cumulative respirometric oxygen uptake data from which the biokinetics were derived. The CO_2 generation rate measurement experiments were performed using both the shaker flask and respirometric reactors to determine the compatibility and reproducibility between the data on CO_2 production in both systems.

Soda lime was used initially as the absorbent for CO_2 evolution studies in the shaker flask and in respirometric reactors, since soda lime served ordinarily as the absorbent of choice in the oxygen uptake respirometric studies for determination of biodegradability and biodegradation kinetic parameters of the priority pollutant organics. Subsequently soda-lime was replaced by KOH as the absorbent in the CO_2 evolution measurement studies, because of the observed slower rate of release of CO_2 from soda-lime during the analysis of CO_2.

Four methods were used for CO_2 measurement in the shaker flask soil slurry systems and in respirometric reactors: (1) use KOH to trap CO_2 released from soda lime, and then titrate KOH using standard HCl before and after CO_2 absorption; (2) use $Ba(OH)_2$ to trap CO_2 released from soda lime, and then titrate $Ba(OH)_2$ using standard HCl before and after CO_2 absorption; (3) use KOH to trap CO_2 released from soda lime, precipitate the $KHCO_3$ and K_2CO_3 formed, using $BaCl_2$ to get $BaCO_3$ which is filtered, dried and weighed; (4) use KOH to trap CO_2, measure pH of KOH solution before and after CO_2 absorption, calculate CO_2 generated according to the changes of pH values using the developed computer program (30). The pH method was shown to be most accurate and provided the most reproducible data. Shaker flask slurry systems data on CO_2 production were generated for the same alkyl phenols used in the respirometric soil slurry reactor studies.

In shaker flask soil slurry systems, incorporating KOH solution as absorbent for the generated CO_2, specially designed flasks with interchangeable KOH traps were incorporated in the study, so that by replacing these traps with fresh KOH solution, continuous CO_2 evolution for the duration of the incubation time can be measured for each shaker flask. The pH method using the computer program was used to calculate the continuous CO_2 generated from each flask in these shaker flask soil slurry systems. Figure 2 illustrates shaker flask absorber assembly designed for determination of carbon dioxide generation during biodegradation in soil slurry systems. This system consists of a flask in which the soil slurry with nutrients and contaminant is mixed. The carbon dioxide generated is quantified by sweeping the headspace of the shaker flask with carbon dioxide free air and absorbing the carbon dioxide generated in potassium hydroxide solution.

In the respirometric reactors, carbon dioxide generation rates were measured utilizing both soda lime and KOH solution as the absorbents of the generated CO_2. The CO_2 evolution studies were undertaken at the same time that oxygen uptake data were generated in the same respirometric reactors for determination of biokinetics of degradations of the alkyl phenols. In experiments in which soda lime was used as absorbent, at appropriate sampling times during the respirometric runs, coincident with designated times of the respirometric oxygen uptake velocity curve, total soda lime amounts were emptied from the reservoirs in the reactors and subjected to the three procedures described for soda lime absobent. The soda lime that was analyzed was replaced with fresh soda lime in the reservoirs for the subsequent CO_2 evolution measurement. This sampling procedure provided CO_2 data for the entire respirometric oxygen uptake experiment. In the experiments using KOH solution as the absorbent, the pH method using the computer program was used to calculate the continuous CO_2 generated from each respirometric reactor in the respirometric soil slurry systems.

Studies on Adsorption/Desorption Equilibria and Kinetics.

Adsorption Studies.   Soil adsorption kinetics and equilibria are measured using batch well-stirred bottles. The soil is initially air dried and then sieved to pass a 2.00 mm sieve. 10 g of soil sample is placed in each bottle and mixed with 100 mL of distilled deionized water containing various concentrations of the compound and mercuric chloride to minimize biodegradation. The soil:solution ratio is expressed as the oven-dry equivalent mass of adsorbent in grams per volume of solution.

The liquid is sampled after 2, 4, 6, 8, 10, 12, 14, 16, 18, 20, and 36 hours. Before the liquid sample can be taken, after the predefined time has elapsed, the bottle contents are centrifuged and the liquid sample is taken using a syringe connected to a 0.45 μm porous silver membrane filter. The filter prevents soil particles from entering the sample.

The concentration of the chemical compound in the liquid sample was analyzed using three methods: (1) standard extraction (EPA method 604 and 610) with methylene chloride followed by GC/MS analysis; (2) HPLC analysis; and (3) scintillation counting of the ^{14}C using radiolabelled compound. All three analytical methods were calibrated using standard solutions of each compound with concentrations levels of 1, 5, 10, 50, 100 and 150 mg/L. The calibration data is used to convert the peak areas (GC/MS or HPLC) or counts of disintegrations per minute (DPM) for radiolabelled compounds to the actual liquid concentration.

HPLC had the distinct advantage of requiring no extraction with a solvent as in the case of GC/MS analysis. However, HPLC method did not have enough sensitivity for liquid phase concentrations below 1 ppm. Solvent extraction was time consuming, used excessive amounts of solvent and often did not yield 100% recovery of the compound. ^{14}C scintillation counting required radiolabelled compound but was fast and easy to use. However, limited number of ^{14}C compounds are available and radiolabelled compounds are expensive. Furthermore, scintillation counting methods result in the generation of radioactive waste that requires proper handling and disposal.

For liquid phase concentrations below 1 ppm, ^{14}C scintillation counting is the preferred method. For higher concentrations, and for compounds with low octanol-water partition coefficient, HPLC is better than extraction with organic solvent and GC/MS analysis. In this study, results from all three analytical methods agreed closely.

From the initial amount of compound and analysis of the liquid phase, the amount of compound adsorbed in the soil is obtained by difference. Equilibrium is defined when the liquid concentration reaches a stationary value, which is usually attained in 24 hours. The equilibrium data are used to obtain Freundlich isotherm parameters.

According to the Freundlich isotherm, adsorption at equilibrium is expressed by

$$X_a/M = S_s = K_d \, S_w^{1/n} \qquad (7)$$

where X_a = Amount of chemical adsorbed in soil (mg)
  M = Mass of soil (Kg)
  K_d = equilibrium constant indicative of adsorptive capacity $(mg/kg)(L/mg)^n$
  S_w = solution concentration at equilibrium after adsorption (mg/L)
  n = constant indicative of adsorption intensity

Desorption Studies.   Desorption studies were conducted by first adsorbing the chemical in the soil until equilibrium is achieved.   Then desorption studies are conducted using two methods: (1) centrifuging the soil slurry to separate the liquid with the soil phase, withdrawing a specified volume of the liquid and adding deionized distilled water; and (2) diluting with deionized distilled water.   In both methods, the concentration in the liquid phase is measured to obtain the desorption kinetics and equilibria.

In the first method, 100 ml of deionized distilled water is mixed with 20 grams of soil and specified concentration of chemical for adsorption. After adsorption equilibrium is attained, the whole solution is centrifuged and 90 ml of supernatant is taken out and replaced with an equal volume of deionized distilled water and with 20 gm/l of mercuric chloride to inhibit biodegradation. Desorpton begins from this time. 20 ml sample is withdrawn at 4, 8, 16, 24, 48, 72, 96 and 120 hours. Each sample is withdrawn from a separate bottle. Each sample is filtered with 0.45 μm filter and extracted with methylene chloride and analyzed using GC/MS technique to determine the concentration of solute.

In the second method, 100 ml of deionized distilled water is mixed with 20 gms of soil and specified concentration of chemical for adsorption.   After adsorption equilibrium is attained, the sample is diluted with an equal volume of deionized distilled water and with 20 gm/l of mercuric chloride to inhibit biodegradation. 20 ml sample is withdrawn at 4, 8, 16, 24, 48, 72, 96 and 120 hours. Each sample is withdrawn from a separate adsorption bottle. The sample was analyzed using extraction with methylene chloride followed by GC/MS analysis. HPLC analysis and ^{14}C scintillation counting analysis for radiolabelled compounds were also used as additional methods to compare the analysis results in desorption studies. Results from all three analytical approaches agreed closely, and subsequently HPLC was used exclusively.

It was found that significant errors resulted when the first method using centrifugation of the soil slurry was used. This was mainly due to difficulties in separating colloidal size particles using the centrifugation, and significant amount of the chemical was adsorbed on these micron size particles. Attempts in using smaller pore size filters failed since significant pressure was necessary to force the liquid through the filter. Hence, the dilution method, which required no separation, was selected in our studies.

Radiochemical Techniques for Determining Biodegradation of Organics in Soil. Radiorespirometric biodegradation protocol was developed for quantitating the biodegradation / mineralization of the organic pollutant compounds in soil slurry systems. The protocol was shown to be particularly applicable to the study of biodegradation of alkyl phenol compounds in the respirometric soil slurry reactors in that it was possible to corroborate the extent of biodegradation data on these compounds as measured by the respirometric oxygen uptake and carbon dioxide evolution in shaker flask systems.

Studies were conducted with phenol and each of the alkyl phenols in concentrations of 50, 100 and 150 mg/L in the respirometric reactors. The phenols were added to the OECD nutrient solution made in deionized distilled water and mixed with either 5, 10 or 15% soil. One ml of ^{14}C fully tagged radiolabeled phenol (concentration equals to 1 μci/ml) was added to each of the experimental and control reactors. One uci equals 2.22×10^6 DPM units (disintegrations per minute). Five ml of 2N KOH solution is used to absorb the $^{14}CO_2$ released from the solution. After appropriate acclimation time depending on the alkyl phenol studied, sampling of KOH in the special holder in the respirometric reactors is initiated by taking out 5 ml KOH volumes with absorbed $^{14}CO_2$ every two hours from each of the reactors. The sampled KOH solution is mixed with a cocktail solution (Ultima Gold) (22) in a 1:10 ratio (1 ml of KOH to 9 ml of cocktail solution) and the sample is then analyzed via liquid scintillation counting technology (Packard TRI-CARB 2500 TR Liquid Scintillation Analyzer). Liquid scintillation efficiency tracing technique (31) is used to quantitate the $^{14}CO_2$ concentration released from the solution with time in the reactor systems.

Liquid scintillation efficiency tracing technique has several advantages over the traditional DPM analysis. These advantages are: (1) no quench curve is needed; (2) the technique can be used for almost all beta and beta gamma emitters; (3) a single unquenched sample is required for the calculation of DPM values; (4) the technique is relatively simple to use; (5) relatively small errors (1-5%) are expected in the DPM calculation; and (6) the results are independant of cocktail density variation, vial size, sample volume, color, chemical quenching and vial composition.

Methodology for Determining Bacterial / Soil Sorption Isotherm. The adsorption isotherm for the bacterial cells is determined by incubating soil microbiota with radiolabelled phenol in a respirometric reactor until an oxygen uptake plateau is obtained, indicating that all phenol has biodegraded either into $^{14}CO_2$, which is adsorbed in the KOH solution and into ^{14}C biomass. The soil suspension is allowed to settle for about 30 minutes. One ml of supernatant is sampled and the ^{14}C activity is measured by liquid scintillation counting. Equilibrium amounts of the ^{14}C biomass adsorbed to the soil is determined by subtracting the ^{14}C present in the biomass in suspension and the ^{14}C present as carbon dioxide absorbed in the KOH solution from the total ^{14}C added initially. The ratio of the biomass adsorbed to the soil and the biomass present in the suspension gives the biomass / soil adsorption isotherm parameter, Kb.

Results and Discussion

Studies on Soil Microcosm Reactors. With the use of these specially designed microcosm reactors, it was possible to acclimate the indigenous microbiota to each class of compounds. Soil samples from the microcosm reactors are used as a source of acclimated microbiota; for measuring oxygen uptake respirometrically to determine biodegradation kinetics; to determine carbon dioxide generation kinetics; and for radiochemical studies to quantitate biodegradation.

Figure 3 shows the cumulative carbon dioxide generation as a function of time for two microcosm reactors, before and after spiking microcosm 1 with a solution of six phenolic compounds and microcosm 4 with OECD nutrients. The cumulative carbon dioxide production increased after spiking with nutrients and with a solution of six phenolic compounds.

Analysis of Oxygen Uptake Data. Respirometric oxygen uptake data were generated for several alkyl phenols using soil slurry systems. This paper reports studies for phenol, p-cresol, resorcinol, catechol and 2,4-dimethyl phenol. Three different compound concentrations; 50, 100, and 150 mg/L and three different soil concentrations, 2, 5, and 10 % were selected for the initial studies. The soil samples were completely mixed with nutrients and compounds, and the oxygen uptake generated was measured respirometrically.

When considering oxygen uptake data for phenol, no significant oxygen uptake occurred when no soil was present due to the absence of soil microorganisms. The total oxygen uptake for 10% soil was higher than for 5% soil, and the lag time for 10% soil was also lower than that for 5% soil.

The oxygen uptake data for 5% soil concentration with 0, 50, 100, and 150 mg/L of phenol concentrations, indicate that at 0% phenol concentration, the oxygen uptake is due to degradation of soil organic matter, and that the oxygen uptake increases with phenol concentration. The oxygen uptake curves did not change appreciably when the soil concentration was increased, except that the lag time decreased for all phenol and alkyl phenol concentrations.

It should be noted that the theoretical oxygen demand at 150 mg/L phenol concentration with no soil was 357.4 mg of oxygen compared to 300 mg/L when soil was present. This indicate sthat some phenol diffused into the soil particle micropores and hence was inaccessible to the bacteria. The sorbed phenol either desorbed very slowly or was irreversibly bound to the soil matrix (32).

Figures 4 and 5 illustrate the representative oxygen uptake curves for phenol at

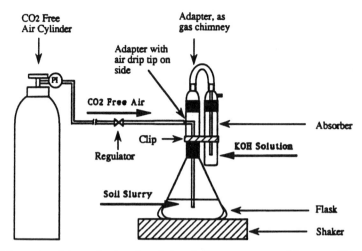

Figure 2.   Schematic showing shaker flask apparatus used for quantifying carbon dioxide generation rates in soil slurry systems.

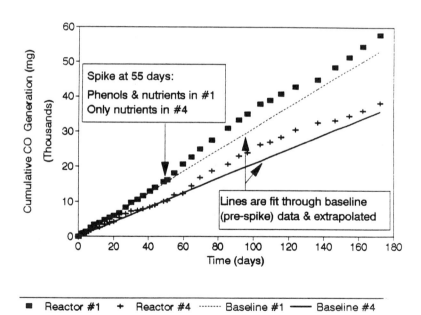

Figure 3.   Cumulative carbon dioxide evolution from two microcosm reactors, before and after spiking microcosm 1 with a solution of six phenolic compounds and microcosm 4 with OECD nutrients.

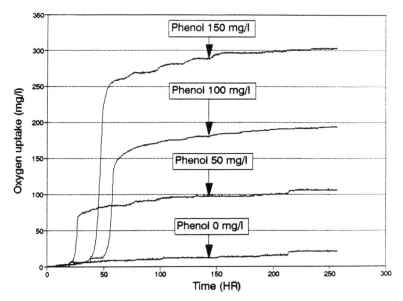

Figure 4.   Respirometric oxygen uptake curves for 5 wt% soil slurry at phenol concentrations of 0, 50, 100 and 150 mg/L.

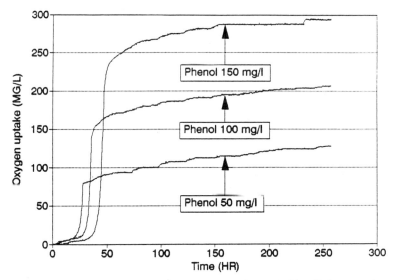

Figure 5.   Respirometric oxygen uptake curves for 10 wt% soil slurry at phenol concentrations of 50, 100 and 150 mg/L.

soil slurry concentrations of 5% and 10% by weight of dry soil respectively and initial phenol concentrations of 0, 50, 100 and 150 mg/L.

Oxygen uptake curves for 5% and 10% soil concentration at various initial concentrations of 2,4 dimethylphenol of 50, 100 and 150 mg/l, are shown in Figures 6 and 7 respectively.

The experimental oxygen uptake data were analyzed and fitted to a non-linear mathematical model comprised of the Monod equation and a nonlinear adsorption isotherm using an adaptive random search technique.

The experimental value of the biomass adsorption parameter, K_b, measured using the procedure discussed earlier is 167.0 (mg/Kg)(L/mg). A high value of K_b indicates that a significant amount of active biomass remains attached to the soil particles as compared to the biomass existing as suspended culture.

The Monod parameters for phenol and alkyl phenols are summarized in Table II. Except for K_s, the other biokinetic parameters do not vary significantly with changes in phenol or soil concentration. The specific growth rate parameter, μ, the biomass yield, Y, and bioproduct yield coefficient, Y_p, did not vary significantly with soil concentration. Variations in K_s and b were due to experimental variability, and errors in curve fitting.

Measurement of CO_2 Generation Rate in Shaker Flask Slurry Systems. Experiments were conducted using shaker flask slurry systems, shown in Figure 2, to determine the amount of carbon dioxide generated due to biodgradation of phenol and alkyl phenols. Four methods for CO_2 measurement were evaluated and the pH method with the developed computer program was shown to be the most accurate and reproducible method. Figures 8 and 9 illustrate the representative curves for the cumulative CO_2 generation for phenol at 0, 50, 100, and 150 mg/L concentration in shaker flask slurry reactors containing 5 and 10% by weight of dry soil, respectively. Figures 10 and 11 show cumulative CO_2 generation data for soil concentration of 5 and 10% respectively and 2,4-dimethylphenol concentrations of 0, 50, 100 and 150 mg/L. The carbon dioxide generation for 0 mg/L of compound is due to soil organic matter. The amount of CO_2 generated increases with compound concentration.

Measurement of CO_2 Generation Rate in Respirometer Reactor Slurry Systems. The relationship of the CO_2 production in shake flask soil slurry and respirometric soil slurry reactor systems was studied through measurement of CO_2 trapped by soda lime in respirometric vessels at various sampling times throughout the course of the respirometric oxygen uptake run. Measurements of cumulative CO_2 generation in the respirometric reactors were made using soda lime as absorbent initially and subsequently with the use of KOH solution which was periodically measured for the pH change.

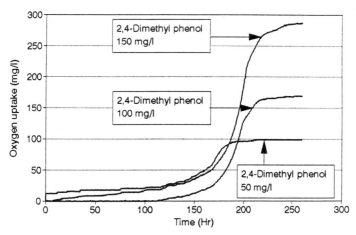

Figure 6.   Respirometric oxygen uptake curves for 5 wt% soil slurry at 2,4-dimethyl phenol concentrations of 50, 100 and 150 mg/L.

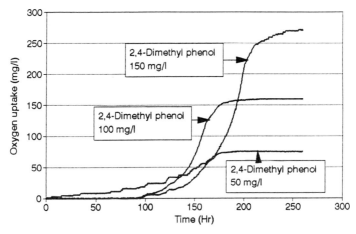

Figure 7.   Respirometric oxygen uptake curves for 10 wt% soil slurry at 2,4-dimethyl phenol concentrations of 50, 100 and 150 mg/L.

Table II. Biokinetic parameters for phenol and alkyl phenols obtained from soil slurry reactor experimental oxygen uptake data

Chemical	μ_w (hr^{-1})	K_w (mg/L)	Y	Y_p	b (hr^{-1})
phenol	0.415	3.47	0.428	6.93	4.38
p-cresol	0.279	8.76	0.476	8.33	7.95
resorcinol	0.360	5.02	0.226	10.10	2.07
catechol	0.560	6.11	0.483	7.40	6.16
2,4-dimethyl phenol	0.492	1.23	0.343	5.90	9.53

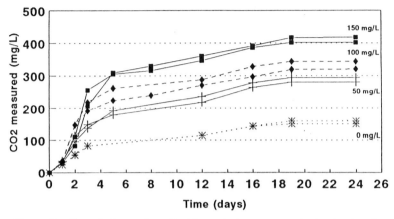

Figure 8. Cumulative carbon dioxide evolution curves, obtained using the shaker flask apparatus (Figure 2), for 5 wt% soil slurry at phenol concentrations of 0, 50, 100, and 150 mg/L.

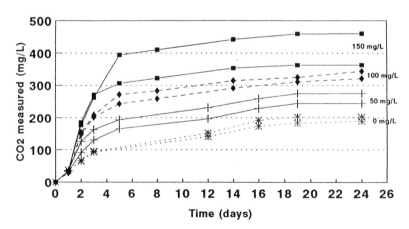

Figure 9. Cumulative carbon dioxide evolution curves, obtained using the shaker flask apparatus (Figure 2), for 10 wt% soil slurry at phenol concentrations of 0, 50, 100, and 150 mg/L.

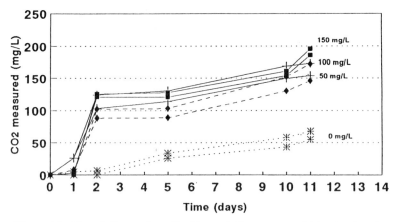

Figure 10. Cumulative carbon dioxide evolution curves, obtained using the shaker flask apparatus (Figure 2), for 5 wt% soil slurry at 2,4-dimethyl phenol concentrations of 0, 50, 100, and 150 mg/L.

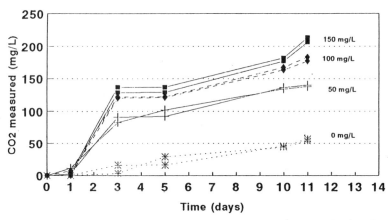

Figure 11. Cumulative carbon dioxide evolution curves, obtained using the shaker flask apparatus (Figure 2), for 5 wt% soil slurry at 2,4-dimethyl phenol concentrations of 0, 50, 100, and 150 mg/L.

Comparisons were made between the cumulative CO_2 concentrations experimentally measured for the different soil and compound concentrations in the shaker flask soil slurry systems and respirometric reactor soil slurry systems measuring oxygen uptake data. Ratio values were developed for the CO_2 experimentally measured in respirometer and shaker flask to the CO_2 calculated from oxygen uptake data.

The compatibility between the data on the CO_2 production in the shaker flask and respirometric vessel tests has been established and verified and the measurement of CO_2 evolution in shaker flask soil slurry systems was shown to be dependable for quantitative CO_2 analysis. The results on the relationship of the two reactor systems provided data on the correlation between the oxygen requirements and CO_2 generation for metabolic activity of microbiota on toxic organics in the soil slurry systems.

The carbon dioxide generation rate provides data on the mineralization rate of the contaminant. The total amount of carbon dioxide generated agrees closely with the appropriate amount of oxygen uptake required to mineralize the compound. This demonstrates that complete mineralization of the compound was achieved in the respirometric experiments.

The quantification of carbon dioxide eneration rates and determination of the mass balance with oxygen uptake data is an important achievement in studying bioremediation of organics in soil systems.

Measurement of $^{14}CO_2$ Generation Rate by Radiorespirometry. Figures 12 and 13 illustrate representative radio-labelled phenol degradation in soil slurry systems spiked with 50, 100 and 150 mg/L of unlabelled phenol for 5 and 10% soil concentration levels, respectively. The DPM data indicate that phenol is preferred over the organic matter normally present in soil. The data demonstrate that measurement of carbon dioxide evolution can be used for determining the kinetics of bioderadation for phenol and alkyl phenols in soil slurry systems.

Furthermore, analysis of the liquid phase showed that little carbon residual remained in the liquid at the end of the experiment sugesting complete mineralization of phenol.

Analysis of Adsorption/Desorption Data. Methodology was developed for the determination of adsorption and desorption equilibria and kinetics for alkyl phenols. Representative adsorption data for phenol and 2,4 dimethyl phenol are shown in Figures 14 and 15, respectively. The corresponding desorption curves for phenol and 2,4-dimethyl phenol are shown in Figures 16 and 17, respectively. Table III provides data on the Freundlich isotherm adsorption and desorption parameters and adsorption intensity coefficients for phenol, 2,4 dimethyl phenol, p-cresol, catechol, hydroquinone and resorcinol.

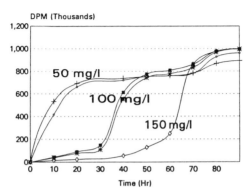

Figure 12. Concentration of radiolabelled carbon dioxide (DPM) absorbed
in KOH solution during biodegradation of phenol using 5 wt% soil
concentration in slurry reactor with 50, 100 and 150 mg/L of
initial phenol concentration.

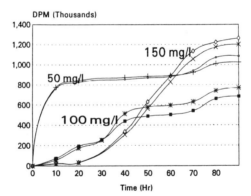

Figure 13. Concentration of radiolabelled carbon dioxide (DPM) absorbed
in KOH solution during biodegradation of phenol using 10 wt% soil
concentration in slurry reactor with 50, 100 and 150 mg/L of
initial phenol concentration.

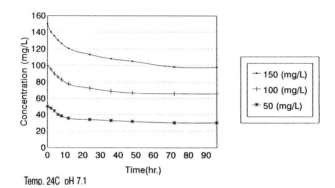

Temp. 24C pH 7.1

Figure 14. Abiotic adsorption curves for phenol in soil slurry system with initial compound concentrations of 50,100 and 150 mg/L.

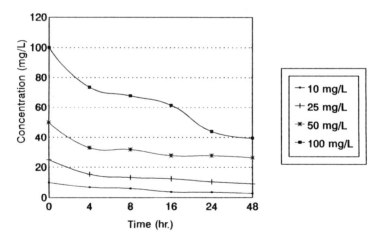

Temp. 24   pH 7.0

Figure 15. Abiotic adsorption curves for 2,4-dimethyl phenol in soil slurry system with initial compound concentrations of 50,100 and 150 mg/L.

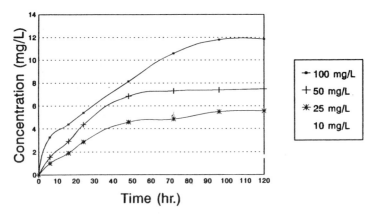

Figure 16. Abiotic desorption curves for phenol in soil slurry system with
initial compound concentrations of 25, 50 and 100 mg/L.

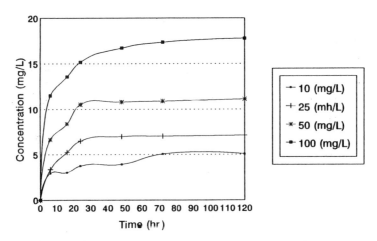

Temp. 24C   pH 7.0

Figure 17. Abiotic desorption curves for 2,4-dimethyl phenol in soil slurry
system with initial compound concentrations of 10, 25, 50 and 100 mg/L.

Table III. Adsorption/Desorption Freundlich Isotherm Parameters for phenol and alkyl phenols in soil

Chemical	Adsorption		Desorption	
	K_d $(mg/Kg)(L/mg)^n$	$1/n$	K_d $(mg/Kg)(L/mg)^n$	$1/n$
phenol	10.50	0.841	12.59	0.774
p-cresol	8.11	0.945	5.98	0.894
2,4-dimethylphenol	15.14	0.768	23.71	0.367
catechol	5.56	0.945	0.19	1.872
hydroquinone	13.43	0.784	23.23	0.212
resorcinol	12.37	0.825	10.02	0.674

The adsorption-desorption kinetics have a direct impact on the compound's biodegradability, as shown by the mathematical model analysis of the oxygen uptake data.

Conclusions

Respirometric studies with soil slurry reactors provides valuable insight into the biodegradation kinetics of compounds in the presence of soil. It has been shown that a Monod kinetic equation in conjunction with a linear adsorption isotherm can provide reliable estimates of the Monod kinetic parameters and the adsorption coefficient. Experiments conducted in our laboratory have demonstrated that cumulative carbon dioxide measurement can be made for soil slurry systems. Carbon dioxide generation in soil slurry systems provides unambiguous measurements of the rate of mineralization of the compound in the presence of soil.

Reconciliation of carbon dioxide generation data with oxygen uptake information is important in determining the biokinetics of not only biotransformation reactions, but also for complete mineralization of the compound.

A protocol developed for quantitative measurement of $^{14}CO_2$ evolution rate by radiorespirometry provides confirmation of mineralization kinetics from carbon dioxide evolution studies and ensures that the net CO2 is generated from mineralization of the compound and not due to natural soil respiration.

Further studies are planned for continuing the respirometric oxygen uptake and carbon dioxide generation measurements for other compounds. This would enable the generation of a database of Monod kinetic parameters and soil adsorption coefficients for various compounds. Eventually, this database will be used to develop predictive models using structure-activity relationships, so that biokinetics of a wide variety of compounds in soil systems can be estimated.

Further studies are also planned for measuring the biokinetics of compounds in compacted soil systems, as opposed to our current measurements in soil slurry systems. This will allow one to determine the impact of soil compaction, oxygen transfer rates, and moisture content on biodegradation.

Attempt will also be made to develop a detailed mathematical model implemented as a computer program, that will use our experimental data and model parameters to aactually quantify rates of bioremediation at contaminated sites. This is significant, since at present, there is no systematic methodology for determining extent of bioremediation at contaminated sites.

Literature Cited

1.   Boethling, R. S., Alexander, M. *Environ. Sci. Technol.* **1979,** 13, 989-991.
2.   Tabak, H. H., Govind, R. Presented at the 4th International IGT Symposium, Colorado Springs, CO, December 1991.
3.   Desai, S., Govind, R., Tabak, H. H., In *Emerging Technologies in Hazardous Waste Management,* Tedder, W., Pohland, F.G., Ed., ACS Symposium Series, 1989.
4.   Tabak, H. H., Desai, S, Govind, R. Proceedings of 44th Annual Industrial Waste Conference, Purdue University, West Lafayette, Indiana, May 9-11, **1989**, pp 405-423 .
5.   Tabak, H. H., Desai, S, Govind, R. In *On-Site Bioreclamation for Xenobiotic and Hydrocarbon Treatment,* Hinchee, R.E., Olfenbuttel, R.R., Ed., Battelle Memorial Institute, Columbus, OH, ButterWorth-Heinemann, **1991**, pp 324-340.
6.   Tabak, H. H., Gao, C., Desai, S., Govind, R. *Water Sci. Tech.* **1992**, 26, No. 3-4, 763-772.
7.   Graves, D. A., Lang, C. A., Leavitt, M. E. *Applied Biochem. Biotechnol.* **1991**, 28/29, 813-826.
8.   Khan, K. A., Krishnan, R., O'Gara, T. F., Missilian, C., Runnells, G. D., Flathman, P. E. Proceedings of the 83rd Annual Air and Waste Management Association Meeting and Exhibition, Pittsburgh, PA, June 24-29, **1990**.
9.   Speitel, G. E, Digiano, F. A. *Wat. Res.* **1988,** 22, No. 7, 829-835.
10. Zelles, L., El-Kabbany, S., Scheunert, I., Korte, F. *Chemosphere* **1989**, 19, No. 10-11, 1721-1727.
11. Scheunert, I., Vockel, D., Schmitzer, J., Korte, F. *Chemosphere* **1987,** 16, No.5, 1031-1041.
12. McDonald, J. P., Baldwin, C., Erickson, L. E. Paper presented at the 4th International IGT Symposium on Gas, Oil, and Environmental Biotechnology, Colarado Springs, CO, December **1991**.
13. Dhawan, S., Fan, L. T., Erickson, L. E., Tuitemwong, P. *Environ. Progress* **1991,** 10, No. 4, 251- 260.
14. Namkoong, W., Loehr, R. C., Malina, J. F. *Hazardous Waste and Hazardous Materials* **1988,** 5, No. 4, 321-328.
15. Dassappa, S. M., Loehr, R. C. *Water Research* **1991,** 25, No. 9, 1121-1130.
16. Special Publication No. 22, 243-269.

17. Scow, K. M., Schmidt, S. K., Alexander, M. *Soil Biol. Biochem.* **1989**, 21, No. 5, 703-708.
18. Weissenfels, Walter D.,Hans-Jurpen Klewer, Joseph Langhoff. *Appl. Micr. Biotechnology* **1992**, 36, 689-696.
19. Robinson, Kevin, G., William S. Farmer, John T. Novak. *Wat. Res.* **1991**, 24, No. 3, 345-350.
20. Weber, Walter T, Paul M. McGinley, Lynn E. Katz. *Water Research* **1991**, 25 No. 5, 499-528.
21. Gordon, Andrew S., Frank J. Millers *Micro. Ecol.* **1985**, 11, 289-298.
22. Loosdrecht Van, Mark C.M., Johannes Lyklema, Willem Norde, Alexander J.B. Zehnder. *Microbiological Reviews* **1990**, 54 No. 1, 75-87.
23. Angley, Joseph T., Mark L. Brusseau, W. Lamar Miller, Joseph J. Delfino. *Env. Sci. Technol.* **1992**, 26, No. 7, 1404-1410.
24. Brusseau, Mark, L., Ron E. Jessup, P. Suresh C. Rao. *Env. Sci. Technol.* **1991**, 25, No. 1, 134-142.
25. Tabak, H. H., L. Lai., X. Yan, C. Gao, I. S. Kim, S. Phanstiel, R. Govind. Presented at the 5th International IGT Symposium on Gas, Oil and Environmental Biotechnology, Chicago, Illinois, September 21-23, 1992.
26. Govind, R., C. Gao, L. Lai, X. Yan, S. Pfanstiel, H. H. Tabak. Presented at the In-Situ and On-Site Bioreclamation, 2nd International Symposium, San Diego, California, April 5-8, 1993.
27. *Methods of Soil Analysis,* Black, C.A., Ed., American Society of Agronomy, Inc., Madison, WI, 1965.
28. OECD, "OECD Guidelines for Testing of Chemicals," EEC Directive 79/831, Annex V, Part C: Methods for Determination of Ecotoxicity, 5.2 Degradation. Biotic Degradation. Manometric Respirometry. Method DGXI, Revision 5 1983.
29. OECD, "OECD Guidelines for Testing of Chemicals," Section 3, Degradation and Accumulation, Method 301C, Ready Biodegradability: Modified MITI Test (I) adopted May 12, 1981 and Method 302C Inherent Biodegradability: Modified MITI Test (II) adopted May 12, 1981, Director of Information, OECD, Paris, France (1981).
30. Mocsny, D. Internal Report prepared for U.S. EPA **1992** University of Cincinnati, Dept. of Chemical Engineering, Cincinnati, OH.
31. Kessler, M.J. *Liquid Scintillation Analysis: Science and Technology,* **1989,** Packard Instrument Company, Meriden, CT, pp. 3-25 - 3-33.
32. Roy, W.R., Krapac, I.G., Chou, S.F.J., Griffin, R.J. *Batch-type Procedures for Estimating Soil Adsorption of Chemicals,* U.S. Environmental Protection Agency, U.S. Government Printing Office: Washington, DC, 1987, EPA/530-SW-87-006-F.

RECEIVED December 6, 1993

Chapter 5

Mobilization of Bi, Cd, Pb, Th, and U Ions from Contaminated Soil and the Influence of Bacteria on the Process

K. W. Tsang[1-3], P. R. Dugan[1], and R. M. Pfister[2]

[1]Idaho National Engineering Laboratory, EG&G Idaho, Inc.,
P.O. Box 1625, Idaho Falls, ID 83415-2203
[2]The Ohio State University, Columbus, OH 43210

Sterile or nonsterile soil experimentally contaminated with bismuth, cadmium, lead, thorium, and uranium as uranyl then incubated with sterile water showed negligible release of metals from either sample. 10 mM cysteine solution mixed with metal-amended soil under each condition: (a) nonsterile; (b) sterile; (c) sterile then inoculated with pure cultures of soil bacterial isolates; indicated that 90.5% Bi, 4.3% Cd, and 25.9% Pb were released from the nonsterile mix within 24 hours. With uranium, 57.9% was released gradually over eight days. The same pattern was observed with sterile soil, but in smaller amounts, 30.6%, 2.6%, 5.2% and 28.3% respectively. Sterile soil containing cysteine then reinoculated with any of four bacteria isolated from the original soil resulted in release of metal greater than from sterile soil. Nonsterile soil released more Bi and U than any of the sterilized reinoculated soils tested. In the case of Cd and Pb some sterilized reinoculated soils released more Cd and Pb than the nonsterile soil while others released less. In all cases extraction of Th was negligible. The results indicated that (a) active microorganisms influence the ability of soil to retain or release metals and (b) cysteine is an effective agent for the release of some metals from soil. 10 mM glycine removed 40.7% and 5.1% of U from nonsterile and sterile soil, respectively. Commercially prepared thioglycollate culture medium resulted in significant release of both Cd and Th from of nonsterile soil.

Background

Among the various environmental concerns, soil and sediment remediation has received considerable attention in recent years because soils and sediments are the

---

[3]Current address: Reno Research Center, U.S. Bureau of Mines, 1605 Evans Avenue, Reno, NV 89512

0097–6156/94/0554–0078$08.00/0
© 1994 American Chemical Society

ultimate repositories for many metals that cycle in the environment as a result of activities such as mining, electroplating, and various manufacturing and industrial processes. There is considerable interest in the remediation of contaminated soils and sediments by so-called soil-cleaning techniques and in the prevention of future contamination via removal of hazardous metals from processing streams prior to deposition into receiving waters. Present methods of metal recovery from waste waters include the use of ion-exchange resins *(1)*, biosorption *(2,3,4)*, chemical precipitation, electrolysis, reverse osmosis and membrane filtration. A variety of chemical technologies may be of value in the extraction of heavy metals from soils and sediments including washing with: water, salts, complexing agents such as ethylenediaminetetraacetate (EDTA) or nitrilotriacetic acid (NTA), mineral acids, strong bases, and some organic acids, e.g. critic acid, that are also complexing agents *(5)*.

An understanding of the complex associations and interactions among soil components (e.g., clay, silt, sand, etc.), the natural microflora (e.g., bacteria and other organisms) and both metal and organic contaminants is fundamental to our ability to accurately model or predict the behavior (bioavailability, toxicity, migration velocity, immobilization, etc.) of contaminants in subsurface environments.

There is a body of evidence indicating that radioactive contaminants in soils are associated primarily with fine soil particles such as clays and microorganisms *(6,7,8)*. It is also known that both clays and microorganisms are capable of concentrating large amounts of metals from solution and that bioconcentration by microorganisms is capable of exceeding both rate and total uptake of certain clays *(7,9)*.

With respect to the behavior of metal contaminants in soil, some indication of relative migration velocities of mixtures of heavy metals (e.g., Bi^{+3}, Cd^{+2}, Pb^{+2}, Th^{+4}, and UO_2^{+2}) in contaminated clay soil can be gleaned from the report of Tsang et al.*(10)*. All of the above ions added as nitrate salts to a clay type soil column strongly sorbed to the soil. The maximum amount of each ion capable of being sorbed by the soil varied as did the relative rate of migration. The rate of migration of these metal ions occurred in the following order: $Cd^{+2} > UO_2^{+2}$, $Pb^{+2} > Bi^{+3}$, Th^{+4}, while the maximum amount of these ions sorbed to the soil was in the reverse order. Bismuth and thorium were not observed to migrate once they were sorbed to the soil. There was also an indication of interaction between Cd^{+2}, the most mobile and UO_2^{+2}, the second most mobile of the ions examined. Continued addition of UO_2^{+2} solution to the soil column resulted in release of Cd^{+2} suggesting that either a common binding site exists for Cd^{+2} and UO_2^{+2} and that there may be a cation exchange occurring or that Cd^{+2} is not tightly sorbed and the equilibrium shifts toward solution in the presence of Cd^{+2} free water. These kinds of interactions and relative migration rates need to be considered when modeling the behavior of metal contaminants in the subsurface environments.

Bioremediation also appears to have value because of its potential economic advantage *(11)*. The purpose of this report is to further demonstrate the role of bacteria as agents to effect the mobilization of hazardous metals from contaminated soil under relatively high and relatively low Eh (i.e., approaching anaerobic conditions). Further, the effectiveness of the amino acid cysteine, a reducing agent as well as a metal complexing agent and a nutrient for many microorganisms *(12)*, for the removal of several hazardous metals from soil is shown in comparison to: (a) a non-reducing but metal complexing amino acid microbial nutrient (glycine) (b) the reducing agent sodium thioglycollate and (c) the presence or absence of microorganisms.

Experimental

Source and characteristics of soil. The soil used was obtained from the Snake River Plain about one mile south of the Radioactive Waste Management Complex (RWMC), the waste storage facility for the Idaho National Engineering Laboratory located between Idaho Falls and Arco, Idaho. The soil sample area has not been used for waste storage and represents soil native to the area.

The surficial deposits in the vicinity of this soil have been previously described as: "eolian and alluvial which range in thickness from less than 0.6 to more than 7.3 m. Alluvial material was transported to the location in times of high runoff and contains gravel. The eolian material probably derived from the finer fraction of alluvial deposits located to the southwest, in the windward direction. Surficial sedimentary materials and interflow sedimentary bed samples have an average median grain size of less than 1 mm with between 70 and 95% material less than 62 μm--predominantly silt and clay sized. The bulk mineralogy indicates that clay and quartz are the dominant minerals. Accessory minerals include potassium feldspar, plagioclase feldspar and a pyroxene. Clay content ranged from 25 to 40% of the samples analyzed and contained 30 to 36% illite, 13 to 26% smectite, 6 to 12% kaolinite, 0 to 26% carbonate and a cation exchange capacity between 11 and 27 milliequivalents per 100 gm soil" *(13)*. The soil moisture content was 14% by weight at the time of sampling. Washed river sand was mixed with the soil in 1:1 ratio by weight to facilitate percolation of the metal solutions used to experimentally contaminate the soil. It has been determined that the sand used in this study did not sorb any of the metals under investigation in significant amounts.

Preparation of metal-amended soil. A 3.5 kg soil/sand mix was placed in a lucite column (76 mm I.D. x 914 mm) equipped with a drain on the bottom. Based on the metal sorption capacity of the soil/sand mix determined previously, the following, in the form of nitrate salts, in a total of 20 l of H_2O, acidified to pH 3 with dilute HNO_3, were added to the column: 3.54 gm Bi; 1.42 gm Cd; 3.54 gm Pb: 3.54 gm Th; 2.83 gm U. The flow rate was measured at 15 ml per min. No attempt was made to alter the natural flow rate of the metal solution. The column was allowed to stand for one week until effluence stopped. The resultant contaminated soil/sand was then mixed thoroughly in a stainless steel container.

The actual metal concentration in the experimentally contaminated soil was determined to establish a baseline. One hundred grams (wet weight) of the metal-amended soil/sand mix was dried to constant weight at 85°C. The dry mix was weighed and then mixed thoroughly. A 1 gm portion (in triplicate) was digested according to procedures described in EPA SW-846 Method 3050 *(14)*. The concentration of Bi, Cd, Pb, Th, and U in the resultant aqueous sample was determined by inductively coupled plasma-atomic emission spectroscopy (ICP-AES). The total amount of each metal in 100 gm (wet weight) of the metal-amended soil/sand mix was then calculated.

Pure Cultures of Bacteria from Metal-Amended Soil

Five hundred milliliters of sterile 10 mM cysteine solution was mixed with 100 gm of metal-amended soil in a 1 L sterile flask. With a sterile pipet, a 1 ml sample was transferred into a test-tube containing 9 ml of sterile trypticase soy broth (TSB) (30 g/L, Becton Dickinson Microbiology Systems, Cockeysville, MD 21030) and mixed thoroughly. A 1 ml aliquot was then transferred into another 9 ml TSB tube and mixed. From this tube 0.1 ml aliquot was dispensed into a sterile petri dish containing ~20 ml of sterile, solidified trypticase soy agar (TSA) (40 g/L) and spread over the entire surface of the agar plate with a sterile L-shaped glass rod. The plate was incubated for 4 d at $22 \pm 3°C$. Bacterial colonies showing different morphologies on the agar plate were transferred individually to separate TSA plates and streaked over the surface of the agar with a sterile inoculating loop. The plates were incubated for 4 d at $22 \pm 3°C$. With proper aseptic techniques, each plate should contain colonies with the same morphology, i.e., pure culture. The pure cultures were subcultured onto tubes of solidified TSA with slanting surface. The inoculated agar slants were numbered #011, 012, 013, etc. and incubated at $22 \pm 3°C$ for 4 d and then kept at 4°C for stock culture.

To inoculate autoclaved sterile soil with a pure culture isolated from the metal-amended soil, 250 ml of TSB was inoculated with a loopful of pure culture from the stock culture and incubated overnight at $22 \pm 3°C$ on a shaker table at 150 rpm. The cells were harvested by centrifugation at 10,000 x g for 10 min. The supernatant was discarded and the pellet was resuspended in 250 ml of sterile phosphate buffer and again centrifuged at 10,000 x g for 10 min. The supernatant was discarded and the pellet of bacterial cells was transferred to the sterile soil.

Soil washing with water. The efficacy of distilled water on removal of metals from soil, either with or without indigenous microorganisms, was evaluated by mixing 200 gm (wet weight) of nonsterile or sterile metal-amended soil with 500 ml of sterile distilled water in a 1 L flask. The mixture was agitated by stirrer motor at 150 rpm for four days at ambient temperature. The content was allowed to settle for 30 minutes. A 6 ml aliquot of the wash water was withdrawn, centrifuged at 3000 x g for 20 min, and then filtered through a 0.45 μm filter disk. A 5 mL sample of the filtrate was diluted to 25 ml with 1% HNO_3 and assayed for Bi, Cd, Pb, Th, and U by ICP-AES.

Effect of cysteine and microorganisms on release of metals from metal-contaminated soil. To investigate the effect of cysteine on the release of metals and the microbial influence on the process, a metal-amended soil/sand sample was pre-washed by filtering 2 L of distilled water through a soil/sand column as previously described.

The pre-washed soil/sand sample was then removed from the column and a 500 ml 10 mM sterile cysteine solution was mixed with 100 gm of the metal-amended soil/sand by stirrer motor at 150 rpm for 8 d under each of the following conditions: (a) nonsterile soil/sand; (b) sterile soil/sand; and (c) autoclaved sterile soil/sand then inoculated with pure cultures of four bacteria that were isolated from the metal-amended soil, designated with an isolate number and later identified as the following organisms: #011, *Arthrobacter oxydans*; #012, *Micrococcus luteus*; #014, *Bacillus megaterium*; and #016, *Arthrobacter aurescens*. Viable microbes were enumerated daily by spread plating 0.1 ml of a series of diluted soil and cysteine mixtures (i.e., 1/10, 1/100, 1/1000, etc.) *(14)* on trypticase soy agar which was incubated for 4 d at 22 $\pm$ 3°C. Eh and pH of the soil/sand and cysteine mixture were measured daily. Eh measurements were accomplished by using a Pt/AgCl redox electrode, and pH was measured with a standard pH electrode. A 6 ml aliquot of the cysteine solution was withdrawn after 30 min settling, centrifuged at 3000 x g for 20 min, and filtered through a 0.45 μm filter disk. Five milliliters of the filtrate was diluted to 25 ml with 1% HNO_3 for metal analysis by ICP-AES.

Effect of glycine and microorganisms on release of metals from metal-contaminated soil. The effect of glycine, an amino acid known to complex some metals and to serve as a nutrient for many microorganisms, was evaluated on release of metals from metal-amended soil. Experimental procedures were similar to that for cysteine extraction, except that 10 mM glycine was substituted for cysteine and only sterile soil and nonsterile soil with indigenous microorganisms were used.

Comparison of sodium thioglycollate solution and nutrient-supplemented sodium thioglycollate medium on release of metals from metal-contaminated soil. The effect of 0.05% sodium thioglycollate solution on release of metals from soil was compared with that of thioglycollate culture medium (29.8 g/L, Difco Laboratory, Detroit, Michigan) which contained bacterial nutrients such as casitone, yeast extract, dextrose, sodium chloride and L-cystine in addition to 0.05% sodium thioglycollate.

One hundred grams of nonsterile metal-amended soil was placed in 500 ml of either solution and allowed to incubate for four days at 22 $\pm$ 3°C with continuous stirring. Metal concentration in solution was assayed as described previously. To ensure the autoclaving process did not change the chemistry and texture of the soil or cause permanent binding of metal ions to the soil particles, a similar study using thioglycollate broth and sterile soil reinoculated with bacteria was performed for comparison.

Results and Discussion

Effect of water washing on release of metals from sterile and nonsterile soils. Soil-washing with distilled water removed only negligible amounts of metals from either sterile or nonsterile metal-amended soil (Table I). The change in pH for nonsterile soil dropped slightly from 7.40 to 7.27 while a slight increase from pH 7.65 to 7.68 was observed with sterile soil suggesting that microorganisms produced a slight

Table I. Metal released from water-washed nonsterile or sterile soil by distilled water after four days with continuous stirring at $22 \pm 3\,^{\circ}C$

		Bi	Cd	Pb	Th	U
Amount of metal (mg) in 200 gm (wet weight) of amended soil/sand mix		152.00	54.00	162.00	86.00	102.00
Nonsterile soil	mg	0.10	0.15	0.05	<0.10	<0.10
	%	0.07	0.28	0.03	<0.12	<0.10
Sterile soil	mg	0.05	0.05	0.05	<0.10	<0.10
	%	0.03	0.09	0.03	<0.12	<0.10

amount of acid during the course of the experiment. Eh changes were from 204 mV to 178 mV for nonsterile soil from 195 mV to 160 mV for sterile soil indicating either that some oxygen was being depleted or that reducing substances were being produced. The ineffectiveness of distilled water in metal removal was expected since distilled water contained no ions for ion-exchange to occur in either soil sample. Furthermore, distilled water contained no nutrients to support the growth and metabolic activities of the microbial population in the nonsterile soil. As a result, there was little change in the chemical and biologic environment to enhance the release of metals from the metal-amended soil.

Effect of cysteine on mobility of metals. In experiments where cysteine was used the largest amount of Bi (137.5 mg/L) was released from nonsterile soil in the presence of 10 mM cysteine solution (Figure 1). The sterile soil only had 46.5 mg Bi/L released. When sterile soil was reinoculated with pure cultures of bacteria the amount of Bi released was intermediate between the nonsterile and the sterile soil (Table II). In all cases, the maximum amount of Bi released occurred within one day. As time proceeded, some of the metal was reabsorbed by the soil. Release of Cd and Pb showed a similar pattern as Bi. However, certain pure cultures demonstrated higher efficacy in release of Cd and Pb than was found in nonsterile soil (Figures 2 and 3). The release of U showed a different trend from Bi, Cd, and Pb. The amount of U released increased gradually with time (Figure 4). A negligible amount of Th was released from the metal-amended soil in all experimental conditions involving cysteine solution. Therefore, the data are not presented in this report.

Variation in pH, Eh and microbe numbers versus time. pH changes from day 1 to day 2 for all experimental conditions varied (Figure 5), but the maximum amount of Bi, Cd, and Pb released was achieved on day 1 and decreased rapidly by day 2. The fluctuating pH changes over the experimental period did not show direct relationship with the gradual release of U under all conditions tested. There was no obvious indication that the amount of uranium released was a result of the pH change and any general relationship of pH and other metals released is inconclusive.

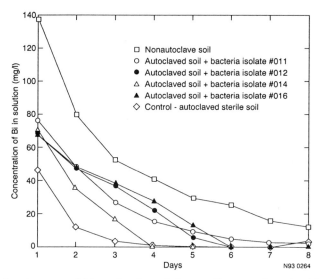

Figure 1. Amount of bismuth released from soil contaminated with 76 mg Bi/100 g (wet wt.) under different experimental conditions versus time at 22 ± 3°C. Isolates: #011, *Arthrobacter oxydans*; #012, *Micrococcus luteus*; #014, *Bacillus megaterium*; and #016, *Arthrobacter aurescens*.

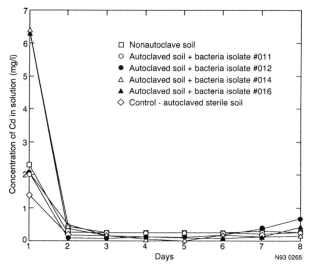

Figure 2. Amount of cadmium released from soil contaminated with 27 mg Cd/100 g (wet wt.) under different experimental conditions versus time at 22 ± 3°C. Isolates: #011, *Arthrobacter oxydans*; #012, *Micrococcus luteus*; #014, *Bacillus megaterium*; and #016, *Arthrobacter aurescens*.

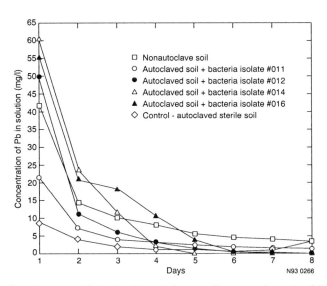

Figure 3.   Amount of lead released from soil contaminated with 81 mg Pb/100 g (wet wt.) under different experimental conditions versus time at 22 ± 3°C. Isolates: #011, *Arthrobacter oxydans*; #012, *Micrococcus luteus*; #014, *Bacillus megaterium*; and #016, *Arthrobacter aurescens*.

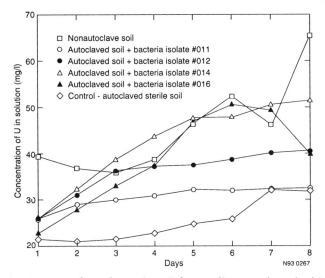

Figure 4.   Amount of uranium released from soil contaminated with 51 mg U/100 g (wet wt.) under different experimental conditions versus time at 22 ± 3°C. Isolates: #011, *Arthrobacter oxydans*; #012, *Micrococcus luteus*; #014, *Bacillus megaterium*; and #016, *Arthrobacter aurescens*.

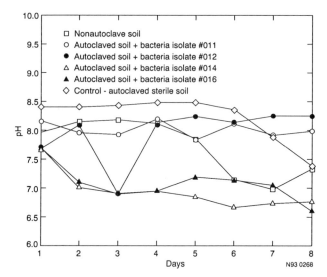

Figure 5.   Change in pH of soil-cysteine mixture under the different experimental conditions presented in Figures 1 through 4. Isolates: #011, *Arthrobacter oxydans*; #012, *Micrococcus luteus*; #014, *Bacillus megaterium*; and #016, *Arthrobacter aurescens*.

Table II. Metal released from previously water-washed soil by 10 mM cysteine under different experimental conditions *

		Bi	Cd	Pb	U
Amount of metal (mg) in 100 gm (wet weight) of amended soil/sand mix		76.00	27.00	81.00	51.00
Control: sterile soil	mg	23.25	0.70	4.25	14.43
	%	30.60	2.60	5.20	28.30
Nonsterile soil	mg	68.75	1.17	21.00	29.54
	%	90.50	4.30	25.90	57.90
Sterile soil + bacterial isolate #011 (*Arthrobacter oxydans*)	mg	38.25	1.03	10.75	14.61
	%	50.30	3.80	13.30	28.20
Sterile soil + bacterial isolate #012 (*Micrococcus luteus*)	mg	33.40	1.05	24.80	18.31
	%	43.90	3.90	30.60	35.90
Sterile soil + bacterial isolate #014 (*Bacillus megaterium*)	mg	34.55	3.20	30.25	23.27
	%	45.50	11.90	37.30	45.60
Sterile soil + bacterial isolate #016 (*Arthrobacter aurescens*)	mg	33.30	3.15	27.60	23.62
	%	43.80	11.70	34.10	46.30

*Five hundred milliliters of 10 mM cysteine solution was added to each soil type. Values represent the metal in solution on the day that showed greatest release of metal over the 8-day period.

Similarly, the inconsistent changes of redox potential in the presence of cysteine (Figure 6) do not appear to correlate to the metal release patterns. However, there was a general trend toward oxidation over the 8-day period. Since the greatest release of most metals occurred in the first 24 hours, it is believed that low potential has a beneficial effect on release of the metals into solution, with the exception of uranium.

Microorganisms in nonsterile soil increased from 2.69 x 10^5 on day 1 to 1.1 x 10^8 colony forming units (CFU)/ml on day 8, while the sterile soil remained sterile through day 3. After three days a gradual increase in population occurred to 5.5 x 10^7 CFU/ml on day 8 due to contamination. Sterility of the soil was confirmed by plating (in triplicate) 0.1 ml of the soil and cysteine mixture on TSA plates and incubated for 4 d at 22 $\pm$ 3°C. The microbial population in sterile soil that was reinoculated with pure culture did not show significant changes in bacteria numbers during the 8-day period (Figure 7), indicating the total microflora

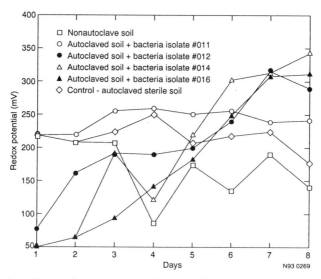

Figure 6. Change in redox potential of soil-cysteine mixture under the different experimental conditions presented in Figure 1 through 4. Isolates: #011, *Arthrobacter oxydans*; #012, *Micrococcus luteus*; #014, *Bacillus megaterium*; and #016, *Arthrobacter aurescens*.

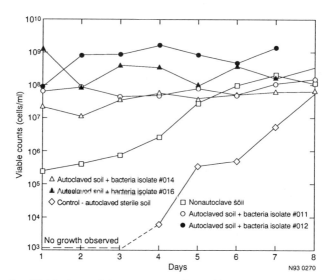

Figure 7. Viable bacterial counts on soil-cysteine mixture under different experimental conditions presented in Figures 1 through 4.

population is not necessarily directly proportional to the amount of metals released from the soil. It appears that there is a threshold value for the microbial population in the conditions tested; further increase in the size of the microbial population did not further increase the amount of metals released.

Cysteine is moderately effective in releasing Bi and to a lesser extent U from sterile soil, but less effective in removal of Cd and Pb. In the presence of microorganisms, the amount of Bi, Cd, Pb, and U released is strongly enhanced. However, a particular microorganism isolate may enhance the release of certain metals but not others. Although the specific mechanism of metal release is not known, several possibilities exist including: (a) indirect mechanisms resulting from metal interactions with microbial metabolites or changes in pH and Eh near the soil colloid without measurable effects on the overall pH and Eh, or (b) alteration of the valence state of metals through oxidation or reduction, or (c) release of metals on decomposition of organic matter that binds metals in the soil.

Effect of glycine on mobility of metals in sterile and nonsterile soil. Extraction of Bi, Cd, Pb and Th ions from sterile and nonsterile soil with 10 mM glycine was similar to that using distilled water; negligible amounts of the metals were released from the soil. For removal of uranyl ions from nonsterile soil, however, glycine was far more effective than water; 20.75 mg was released from soil as compared to 0.10 mg (Table III). It appears that the release of uranyl ions from the soil was enhanced by the microbial activity since a substantially smaller amount of uranyl ions (2.62 mg) was released in the case where sterile soil was used. Redox potentials measured during the 8-day period fluctuated between 76 mV and 167 mV for nonsterile soil, and between 137 mV and 148 mV for sterile soil. pH fluctuated between 6.76 and 8.14 for nonsterile soil and between 8.38 and 8.50 for sterile soil. Viable bacterial counts indicated the sterile soil remained sterile through day 6 and increased to 1.08×10^4 by day 8, whereas the nonsterile soil maintained a relatively stable population (8.50×10^8 - 1.45×10^9) throughout the study.

Table III. Metal released from previously water-washed nonsterile or sterile soil by 10 mM glycine *

		Bi	Cd	Pb	Th	U
Amount of metal (mg) in 100 gm (wet weight) of soil/sand mix		76.00	27.00	81.00	86.00	51.00
Nonsterile soil	mg	0.07	0.43	0.18	0.57	20.75
	%	0.09	1.60	0.22	0.66	40.69
Sterile soil	mg	0.03	0.41	0.05	0.03	2.62
	%	0.04	1.52	0.06	0.03	5.14

*Values represent the metal in solution on the day that showed greatest release of metal over the 8-day period after addition of 500 ml of 10 mM glycine solution.

Effect of thioglycollate on mobility of metals in sterile and nonsterile soil. Removal of metal ions from soil using 0.05% sodium thioglycollate showed a small amount of U released from the soil (2.54 mg U), while other metals released were negligible (Table IV). Thioglycollate medium removed lesser amount of Bi, Pb, and U than 10 mM cysteine solution, but was able to remove significantly larger amounts of Cd and Th from the soil (Table V).

Table IV. Metal released from previously water-washed nonsterile soil by sodium thioglycollate *

		Bi	Cd	Pb	Th	U
Amount of metal (mg) in 100 gm (wet weight) of soil/sand mix		76.00	27.00	81.00	86.00	51.00
Nonsterile soil	mg	0.024	0.086	0.035	0.029	2.540
	%	0.032	0.319	0.043	0.034	4.980

*Five hundred milliliters of 0.05% sodium thioglycollate was added to 100 gm of soil and incubated at 22 $\pm$ 3°C for four days with continuous stirring.

Table V. Metal released from previously water-washed nonsterile soil or sterile soil reinoculated with microorganisms in the presence of thioglycollate medium *

		Bi	Cd	Pb	Th	U
Amount of metal (mg) in 100 gm (wet weight) of soil/sand mix		76.00	27.00	81.00	86.00	51.00
Nonsterile soil	mg	28.00	10.00	9.40	15.10	18.68
	%	36.84	37.04	11.60	17.56	36.63
Nonsterile soil	mg	32.25	4.52	10.35	9.40	14.93
	%	42.43	16.74	12.78	10.93	29.27

*Five hundred milliliters of thioglycollate medium (29.8 g/L, Difco Laboratory, Detroit, Michigan) was added to 100 gm of soil and incubated at 22 $\pm$ 3°C for four days with continuous stirring.

The difference in metal extraction capability of sodium thioglycollate and thioglycollate medium may be due to the direct or indirect influence of the additional nutrients present in the commercially prepared culture medium or metabolites such as organic acids produced.

Sterile soil that was reinoculated with microbes showed comparable results to the nonsterile counterpart in release of metals in thioglycollate medium, indicating that the autoclaving sterilization did not cause the permanent binding of Bi, Cd, Pb, Th, and U ions to soil particles.

Conclusions

o   The presence of microorganisms in soil contaminated with Bi, Cd, Pb, Th, and UO_2 ions strongly influences the mobilization of those ions.

o   Cysteine, a reducing as well as a metal complexing agent and a nutrient for many microorganisms was moderately effective in releasing Bi and to a lesser extent U from sterile soil, but less effective in release of Cd, Pb, and Th. Mobilization of Bi, Pb, and U in the presence of cysteine was strongly enhanced in the presence of microorganisms. However, a single bacterial isolate may enhance the release of certain metals but not others.

o   Sodium thioglycollate, a reducing agent, when added to nonsterile soil, resulted in a small release of U from soil and lesser amounts of Bi, Cd, Pb, and Th. When supplemented with other microbial nutrients, the mixture resulted in release of larger amounts of Bi, Cd, Pb, Th, and U.

o   Glycine, a metal complexing amino acid that is also a nutrient for many organisms was ineffective in removal of Bi, Cd, Pb, and Th from sterile or nonsterile soil. Removal of U by glycine, however, was enhanced in the presence of microorganisms.

o   Although the specific biological mechanism(s) of metal release from contaminated soil during these experiments is not known, several possibilities exist. Enhancement of metal release from soil by microorganisms may be due to:

1.   Indirect mechanisms resulting from metal interactions with microbial metabolites or changes in pH and/or Eh near the soil colloid. Such changes in pH and/or Eh may not measurably affect the overall pH and Eh.

2.   Alteration of the valence state of metals through oxidation or reduction.

3.   Release of metals due to decomposition of organic matter in soil.

o   The combined use of metal sorption by a clay soil column reported previously *(10)* and subsequent extraction of metals by cysteine in the presence of active microbial populations offers a potential process for separation of some radionuclides from nonradioactive metal ions in waste water or in contaminated soils or for removal of some radioactive metal contaminants from solid surfaces, e.g., from contaminated equipment. It also offers the possibility for concentrating radioactive contaminants thereby decreasing the volume of contaminated soil that will require long term storage.

Acknowledgment

Prepared for the U.S. Department of Energy through the EG&G Idaho LDRD Program under DOE Field Office, Idaho Contract DE-AC07-76IDO1570.

Literature Cited

1.   Korkisch, J. *Handbook of Ion Exchange Resins: Their Application to Inorganic Analytical Chemistry.* CRC Press, Inc. Boca Raton, FL. **1989**. *Vol. 1.*

2.   Burke, B. E.; Tsang, K. W.; and Pfister, R. M. "Cadmium sorption by bacteria and freshwater sediment." *Journal of Industrial Microbiology* **1991**, *Vol. 8*, pp. 201-208.

3.   Mullen, M. D.; Wolf, D. C.; Ferris, F. G.; Beveridge, T.; Flemming, C. A.; and Bailey, G. W. "Bacterial sorption of heavy metals." *Applied and Environmental Microbiology* **1989**, *Vol. 55*, pp. 3143-3149.

4.   Dugan P. R. "The function of microbial polysaccharides in bioflocculation and biosorption of mineral ions." In *Proceedings Symp. on Flocculation in Biotechnology and Separation Systems.* Editor, Y. Attia; Elsevier, Amsterdam, The Netherlands, **1986**; pp. 337-350.

5.   *Technology approaches to the cleanup of radiologically contaminated superfund sites.* EPA/540/2-88/002. U.S. Environmental Protection Agency. Washington, DC; **1988**.

6.   Borrowman, S. R. and Brooks, P. S. "Radium removal from uranium ores and mill tailings." Rept. U.S. Bureau of Mines #8099. **1975**.

7.   Pfister, R. M.; Tsang K. W.; and Dugan P. R. "Biological and mechanical release of heavy metal contaminants in soil." *Proceedings of Bioprocess Engineering Symposium*, American Society Mechanical Engineers; **1990**, pp. 69-74.

8.   Weber, W. J.; McGinley, P. M.; and Katz, L. E. "Sorption phenomena in subsurface systems: Concepts, models and effects on contaminant fate and transport." *Water Research* **1991**, *Vol. 25*, pp. 499 428.

9.   Titus, J. A. and Pfister, R. M. "Effects of pH, temperature and Eh on the uptake of Cd by bacteria and an artificial sediment." Bull. Environmental Contamination Toxicology **1982**. *Vol. 28*, pp. 697-704.

10.   Tsang, K. W.; Dugan, P. R.; and Pfister, R. M. "Removal of bismuth, cadmium, lead, thorium, and uranyl ions from contaminated water by filtration through clay soil" In *Proceedings International Topical Meeting in Nuclear and Hazardous Waste Management.* Spectrum '92, American Nuclear Society; **1992**, pp. 1177-1182

11.   Lundgren, D. C. "Biotic and abiotic release of inorganic substances exploited by bacteria." *Bacteria in Nature* **1989**, *Vol. 3*, pp. 293.

12.   Hanson, W. C.   "Transuranic elements in the environment."   U.S. Department of Energy (DOE/TIC-22800), Technical Information Center, **1980**.

13.   Rightmire C. T. and Lewis, B. D. "Hydrogeology and geochemistry of the unsaturated zone.   Radioactive Waste Management Complex."   Idaho National Engineering Laboratory, U.S. Geological Survey Report. 8704198, **1987**.

14.   *EPA Test Methods for Evaluating Solid Waste*.   Volume IC: Laboratory Manual, Physical/Chemical Methods.   **1986**, SW-846, Third edition.

15.   Koch, A. L. "Growth Measurement."   In *Manual of Methods for General Bacteriology*.   American Society for Microbiology.   Washington, DC, 1981. pp. 179-207.

RECEIVED January 14, 1994

WASTE MINIMIZATION
AND MANAGEMENT TECHNOLOGIES

Chapter 6

Artificial Intelligence Approach to Synthesis of a Process for Waste Minimization

T. F. Edgar[1] and Y. L. Huang[2]

[1]Department of Chemical Engineering, The University of Texas at Austin, Austin, TX 78712
[2]Department of Chemical Engineering, Wayne State Unversity, Detroit, MI 48202

One of the most effective ways of minimizing wastes from their sources is to design an environmentally clean process which provides a satisfactory level of controllability. In the present work, a systematic module-based synthesis approach is developed to design such a process with minimal waste generation. This approach is featured by adding the dimension of structural controllability to the conventional capital and operating cost functions, and elaborating waste minimization strategies as constraints. Due to insufficient information and incomplete data regarding the mechanism of waste generation at the process design step, artificial intelligence techniques are utilized to represent waste minimization strategies and evaluate structural controllability. The efficacy of the proposed approach to waste minimization is illustrated by synthesizing a cost-effective and highly controllable process capable of minimizing phenol containing waste streams in an oil refinery.

Rapidly changing industrial technologies are often accompanied by the increased generation of various hazardous or toxic wastes. The major portion of the wastes is mainly from chemical plant and petroleum refineries. These wastes, if improperly dealt with, can threaten both public health and environment. Consequently, waste minimization and management (WMM) are a major concern of process engineers (1-2).

The EPA has established a waste priority hierarchy beginning in 1976 to promote better WMM alternatives. During last two decades, most of the effort on WMM has focused on waste incineration/treatment and land disposal. These are now the lowest priorities in EPA's hierarchy, due to the current emphasis on source reduction or in-plant pollution prevention. Recently, it has been

0097–6156/94/0554–0096$08.00/0
© 1994 American Chemical Society

recognized that the production of wastes from a chemical or petrochemical process is a function of process design and the manner in which the process is controlled and operated *(3-7)*.

To effectively realize waste minimization (WM), it is beneficial to examine the basic characteristics of WM engineering in the process industry (Huang and Fan, *Intl. J. Computers in Industry,* in press).

(1)   WM is a multi-disciplinary area involving engineering, chemistry, biology, fluid mechanics, mathematics, statistics, economics, and regulations; thereby being a knowledge-intensive area. The acquisition and representation of the knowledge from the experts in these diverse fields is the focal point for successful WM.

(2)   WM is heavily dependent on expertise. The behavior of a process generating wastes cannot be easily described by rigorous mathematical models. Qualitative analysis of the process is thus always necessary.

(3)   The available information pertaining to WM is frequently uncertain, imprecise, incomplete, and qualitative in the design stage. Hence, standard mathematical tools may be incapable of dealing with it.

(4)   A large number of regulations and strategies for WM can be expressed as rules. The symbolic knowledge in the rules has to be appropriately represented and manipulated.

Under these circumstances, it is very difficult, if not impossible, to resort to conventional algorithmic methods to design a process with minimal waste generation. On the other hand, artificial intelligence techniques such as knowledge-based approach and fuzzy logic are viable alternatives. The knowledge-based approach is powerful in acquiring both structured and unstructured symbolic knowledge, and efficient in representing and manipulating them *(8)*. Today, numerical computation approaches can be embedded into a knowledge-based system to form a hybrid system. Fuzzy logic is capable of dealing with structured numerical knowledge and imprecise information, and provides a way to interpolate between regions with different rules *(9)*. A methodology incorporating the knowledge-based approach and fuzzy logic is thus highly advantageous.

Design Philosophy

With the recognition of the above basic characteristics of WM engineering, a design approach should have the following features:

(1)   Early incorporation of the WM strategy. Process design consists of three major phases in sequence: synthesis, analysis/optimization, and detailed design; among them, process synthesis is the most abstract, therefore the most difficult phase. Clearly, the implementation of WM strategy in the later phases in process design, especially in the detailed design phase, is very limited, because waste generation caused by an improper process structure is extremely difficult to manage. It is desirable, therefore, to incorporate WM strategies starting from the earliest stage in process design, namely the preanalysis stage in process synthesis.

(2)   Generic rather than problem specific. The design approach should be general enough to easily accommodate different restrictions on waste generation. These restrictions may be on the generation of waste energy as well as waste species in different forms, such as gaseous, liquid, or solid. In an oil refinery, for instance, five major sources of generating hazardous or toxic wastes have been identified: process systems, power plant, storage and handling, waste water treatment, and miscellaneous *(10)*. Among them, wastes from process systems are the most difficult to deal with due to complexity. When crude oil enters an oil refinery, it is a heterogeneous mixture of hydrocarbons and various impurities. Widely differing types of crude require different refining techniques and yield different product mixes. As by-products, a large amount of pollutants are generated from various processes. To minimize the generation of pollutants of these types, a general design approach is highly desirable.

In-Plant Waste Minimization Strategy

Even if a process is specifically designed for WM, the waste generated from it may still exceed a tolerable limit during its operation, due to not considering the effect of process structure on WM. Most processes experience disturbances with different intensities during operations. These disturbances can propagate in a process if disturbance propagation paths exist. Intense disturbance propagation inherited in the process structure may make tight control of the concentrations of hazardous species unattainable. Consequently, we need to synthesize a process which is not only cost-effective, but also highly controllable in terms of the quantity and toxicity of waste streams generated.

Representation of Waste Minimization Strategies. The regulations and process designers' experience for WM are usually expressed in rule form. Various rules are available elsewhere *(2, 7, 11, 12)*. These rules can be classified into the sets for: (i) measuring the toxicity of species, (ii) selecting a method for separating hazardous or toxic species, (iii) determining a separation sequence of stream components, (iv) evaluating the feasibility of recycling hazardous or toxic species, (v) selecting the strategies for recovering waste energy, (vi) implementing minimum capital and operating costs, (vii) enhancing structural controllability, (viii) modifying a synthesized process structure, and (ix) making a trade-off among capital cost, operating cost, and WM.

While these rules are expressed in IF-THEN rule form, they need to be quantitatively represented. For instance, one rule is:

IF   a process stream to a reactor contains hazardous species,
THEN   the species should be separated first before the stream enters the reactor in order to avoid the deactivation of catalyst in it.

This is a crisp rule in which no fuzziness is involved. Variables y_1 and y_2 are introduced to represent the species in a stream and the separation sequence of the hazardous species, respectively. Moreover, crisp sets A_1 and A_2 are defined as the set of hazardous species and that of hazardous species to be separated first, respectively. Two crisp membership functions can be defined below.

$$\mu_{A_i}(y_i) = \begin{cases} 1 & \text{if } y_i \in A_i \\ 0 & \text{if } y_i \notin A_i \end{cases} \quad i=1,2 \qquad (1)$$

Thus, the rule has the following two-valued logic expression.

$$A_1 \longrightarrow A_2. \qquad (2)$$

However, many other rules are fuzzy in nature as illustrated below.

IF a process stream experiences severe disturbance of mass flowrate at its inlet, AND the concentration of a species at the outlet of another stream must be controlled precisely to prevent pollutant generation,

THEN these two streams should not be matched in an extractor.

Since the imprecise information is involved in both premise and consequence of the rule, fuzzy set theory should be employed. Fuzzy variables z_1, z_2, and z_3 can be defined as disturbance of mass flowrate of a stream, the fluctuation of the concentration of a species at the outlet of another stream, and the preference of matching these two streams, respectively. Correspondingly, three fuzzy sets, $B_{1,1}$, $B_{1,2}$, and $B_{1,3}$, may be introduced to represent the concepts of severe, moderate, and slight disturbances, respectively.

$$\mu_{B_{1,1}}(z_1) = \begin{cases} 1 - 25z_1, & 0 \le z_1 \le M_1 \\ 0, & z_1 > M_1 \end{cases} \qquad (3)$$

$$\mu_{B_{1,2}}(z_1) = \begin{cases} 25z_1, & 0 \le z_1 \le M_1 \\ 3 - 50z_1, & M_1 < z_1 \le M_u \\ 0, & z_1 > M_u \end{cases} \qquad (4)$$

$$\mu_{B_{1,3}}(z_1) = \begin{cases} 0, & 0 \le z_1 \le M_1 \\ 50z_1 - 2, & M_1 < z_1 \le M_u \\ 1, & z_1 > M_u \end{cases} \qquad (5)$$

Similarly, fuzzy sets $B_{2,1}$, $B_{2,2}$, and $B_{2,3}$ should be defined to represent the concepts of high, moderate, and low control precision of species concentration, respectively. Fuzzy sets $B_{3,1}$ and $B_{3,2}$ must be designated according to the concepts of preferred and not preferred stream match, respectively. The fuzzy rule can thus be expressed by the following fuzzy logic form.

$$\wedge \{B_{i,j} \mid i=1,2; \ j=1,2,3\} \longrightarrow \wedge \{B_{3,k} \mid k=1,2\} \qquad (6)$$

Classification of Process Information. The representation of WM strategies, as described in the preceding section, indicates that process information directly influencing WM must be carefully classified. This includes the disturbances of temperature, pressure, mass flowrate, and species concentration at the inlets of process streams, as well as the tolerable ranges of the fluctuation of concentrations of hazardous species at the outlets of process streams. In a separation process, for instance, for each stream i, a disturbance of source concentration of waste species p, $\delta y_{p_i}^s$ in both positive and negative directions, and that of mass flowrate, δM_i^s, also in both directions, lead to a change in mass flowrate of the species p, δM_{p_i}:

$$\delta M_{p_i} \propto \max \{ |M_i \delta y_{p_i}^{s(+)} - \delta M_i^{(+)} (y_{p_i}^t - y_{p_i}^s)|,$$
$$|M_i \delta y_{p_i}^{s(-)} - \delta M_i^{(-)} (y_{p_i}^t - y_{p_i}^s)|\} \qquad (7)$$

This example shows that different types of disturbance variables may be lumped into a single variable, which simplifies the quantification of the overall influence of all related disturbances.

Generally, the more intense the disturbances in the input variables, the greater the deviation in the output variables from their normal values. Based on their magnitudes, the disturbances can be classified into a number of degrees, such as *very slight, slight, moderate, severe,* and *very severe* disturbances. The degrees are quantified by fuzzy sets as illustrated in Figure 1.

Usually, it is unnecessary to control all output variables of a process to the same level of precision. For example, the composition of a highly toxic species of a stream must be controlled very tightly, while that of a non-toxic component of a stream is less critical. Based on the complexity of a WM problem, the control precision of the output variables can be divided into several grades, such as *very low, low, moderate, high,* and *very high.* The quantification of levels of control precision is similar to that described above.

A disturbance at the inlet of a stream must propagate to the outlet of another stream if a downstream path exists *(13)*. When the two streams match directly through a process unit, the effect of this disturbance on the outlet of another stream is quick and drastic. When they match indirectly through two process units, such an effect is slower and moderate. When the disturbance propagates through many process units before reaching an outlet, its effect dissipates. The more the number of process units involved in disturbance propagation, the greater the extent of dissipation of disturbance influences. The patterns of disturbance propagation can be defined based on the type of a process system. In an exchanger network, four patterns *(very severe, severe, moderate,* and *negligible* propagation) can be introduced, depending on the number of process units through which a disturbance travels *(14)*.

The patterns of disturbance propagation in a process can be illustrated by the example in Figure 2. This mass exchanger network (MEN) contains four mass-exchange-based process units (e.g., an extractor, an absorber, or an adsorber for recovering hazardous species) as illustrated by the grid diagram of Figure 2a. In this figure, the degree of a disturbance caused by the fluctuations of concentration and/or mass flowrate is indicated by the number of solid circles, "•"s. For instance, symbol "••" represents moderate disturbance. The

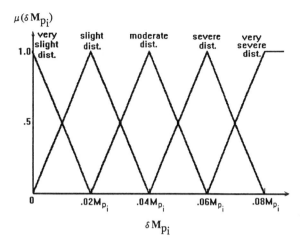

Figure 1. Quantitative representation of fuzzy linguistic terms, VERY SLIGHT, SLIGHT, MODERATE, SEVERE, and VERY SEVERE DISTURBANCES.

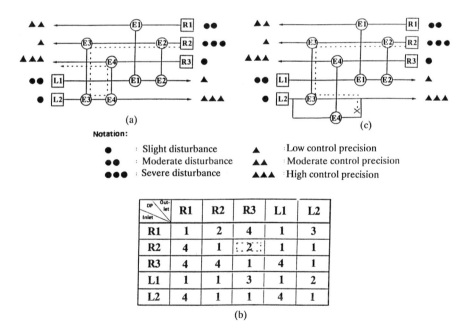

Figure 2. Disturbance propagation through a mass exchanger network and modification of the network structure.

precision level of control at a stream outlet is indicated by the number of triangles, "▲"s. For example, the symbol "▲▲▲" represents high control precision.

All patterns of disturbance propagation involved are listed in the disturbance propagation table of Figure 2b. The first column of the table designates the inlets of the five streams; the first row designates their outlets. Each integer in the remaining entries of the table represents the pattern of disturbance propagation. For example, a disturbance exerted at the inlet of rich stream R_2 propagates through process units 3 and 4, and reaches the outlet of rich stream R_3. In this disturbance propagation path, two process units are involved. According to the definition, this is severe propagation (pattern-2), and thus an integer of 2 is assigned to entry (R_2, R_3). This disturbance propagation is undesirable since the species concentration at the outlet of rich stream R_3 should be controlled very precisely, but it is constantly disturbed by a severe disturbance from rich stream R_3. This WM problem due to improper process structure can be resolved in a number of ways. The process modification in Figure 2c is one possible solution if it is thermodynamically feasible. In this solution, the stream splitting terminates the disturbance path from rich streams R_2 to R_3 with a negligible increase of capital cost and no increase of operating cost.

Structural Controllability. The degree of structural controllability of a process reaches a maximum when the occurrences of undesirable disturbance propagation and their severities are at a minimum, thereby minimizing the possibility of waste generation due to the process structure. Thus, the structural controllability can be assessed by examining the patterns of disturbance propagation through the process. To evaluate it quantitatively, it is convenient to define a disturbance vector D, control precision vector C, and disturbance propagation matrix P for a process having N streams. D comprises all existing disturbances which are detrimental to WM. Each element, d_i, in D represents the intensity of a disturbance exerted at the inlet of stream i. C specifies the levels of control precision required for all output variables. Each element, c_j, in C represents the control precision required at the outlet of stream j. P lists all disturbance propagation in the process. The value of element $p_{i,j}$ in P corresponds to the intensity of the propagation. Each element in D, C, and P is quantitatively evaluated by fuzzy set theory as discussed previously.

To facilitate the evaluation of structural controllability in terms of WM, and to compare various process flowsheets, the index, I_{sc}, is created which has the following general form.

$$I_{sc} = \frac{E_{tot,max}(D,C,P) - E_{tot}(D,C,P)}{E_{tot,max}(D,C,P) - E_{tot,min}(D,C,P)} \qquad (8)$$

where $E_{tot,max}(D,C,P)$ and $E_{tot,min}(D,C,P)$ characterize a process with maximum disturbance propagation and the one with minimum disturbance propagation, respectively. Consequently, the former is the least controllable with the maximum generation of waste which is definitely undesirable; the latter is the most controllable with the minimum generation of waste which is also undesirable due to extremely high capital and operating costs. In fact, to completely eliminate waste generation in a process plant is usually unreachable. Our goal is to design a highly controllable process which is characterized by the terms, $E_{tot}(D,C,P)$. For a separation process using extractors, absorber, and distillation columns or an energy recovery process using heat exchangers, the index can be found in *(14)*.

Module-Based Synthesis

For the knowledge extracted and formalized thus far to be practical, it is imperative that a stagewise procedure for the synthesis be developed; in other words, the knowledge represented in the preceding sections should be systematically organized. Following the three stages in process synthesis, the synthesis procedure consists of three major modules, associated with a number of additional modules to perform specific tasks. The relationships of these five modules are illustrated in Figure 3.

Preanalysis Module. The major function of the preanalysis module includes the estimation of both of capital and operating costs and making decisions on stream matching which should lead to the least waste generation. The costs are estimated by pinch technology *(11)*. In the present work, capital cost is approximated by counting the total number of process units, a heuristic which is commonly used in process synthesis.

For a synthesis problem, process input and output variables are to be identified; process data including normal operation point and fluctuations are analyzed and classified by the approach in the preceding sections. After classification, a waste minimization assessment table is constructed by activating the waste minimization enhancement module. The values of certain grids, i.e., the values of the element p_{ij} in the disturbance propagation matrix P, must be pre-assigned in the table according to the rules generated for implementing WM strategies. Such a WMA table is termed the admissible WM table. The pre-assignment of values to these grids implies the most favorable and the least favorable decisions on placement of process units.

Structure Invention Module. A series of decisions is made on the selection and placement of process units to match pairs of process streams. Each decision's effect on the WM is assessed through evaluation of the index of structural controllability, I_{sc}. Note that each stream match directly introduces a disturbance propagation path, which may influence WM. The recommendations of the locations of process units to be placed, which are reflected in the admissible WM table, should be adopted gradually. Several sets of rules for reducing the total cost and improving WM are applied to ensure the identification of an optimal solution. This module needs to repeatedly activate the waste minimization enhancement module and the stream matching module.

Structure Evolution Module. The resultant process flowsheet is examined by the structure evolution module. Usually, when a number of separately synthesized sub-systems in the structure invention module are combined to form a complete process, two undesirable situations may occur: (1) the total number of process units exceed the minimum requirement, and (2) the WM may deteriorate due to newly introduced intense disturbance propagations. Hence, trade-offs need to be made among the number of process units, energy or material consumption, and WM characterized by structural controllability. The trade-off is usually based on engineering judgment. The preference in this work is given to the prevention of intense disturbance propagation to the process streams whose outlet concentrations of hazardous species need be controlled very precisely.

Waste Minimization Enhancement Module. In this module, the admissible WM table needs to be constructed to impose restrictions on stream matching to prevent undesirable disturbance propagation. The main part of the table is disturbance propagation matrix P as discussed in the preceding section. In the table, some elements of P have preassigned values corresponding to either preferred or disallowed stream matches. Note that at the preanalysis stage, it is impossible to assign values to all elements in the table because of the lack of knowledge about the detailed interconnections among streams through process units.

In the procedure, a value of "1" of the element, $p_{i,i}$, represents the unavoidable very severe propagation (pattern-1) originating from the inlet of stream i to its own outlet. A value of "1" of the element, $p_{i,j}$, $i \neq j$, implies a direct match between streams i and j, which yields two pattern-1 disturbance propagation. This assignment is made according to the intense-propagation-diverting rule and minor-propagation-introducing rule. Note that the match introduced should be thermodynamically feasible. This type of match is useful for diverting severe disturbance propagation to a stream whose hazardous species concentration needs to be controlled precisely. Thus, waste generation can be effectively restricted.

A value of "0" to the elements, $p_{i,j}$, $i \neq j$, implies that an indirect match will lead to a downstream path spanning at least four process units; this is negligible propagation (pattern-4). This assignment represents the complete or almost complete isolation between streams i and j. Thus, a disturbance of stream i can not propagate to the outlet of stream j. This is, in fact, a type of strict constraint on waste generation.

The values of "0.5" and "0.25" represents the restrictions on the path length of disturbance propagation between a pair of streams. The value of "0.5" indicates that the two streams are connected through at least two process units; it corresponds to pattern-2 propagation. The value of "0.25" implies that the two streams are linked through at least three process units; it corresponds to moderate propagation (pattern-3). The assignment of these different values to certain elements in matrix P also enhances WM through controlling disturbance propagation.

Stream Matching Module. The selection of a stream match from a set of match candidates is always based on the information provided by the WMA and the admissible WM tables. Each match candidate needs to be evaluated according to fuzzified heuristic rules. A candidate with the highest priority is always selected. Although it is problem specific, a general procedure for applying the heuristic rules is developed.

Application

The approach has been successfully applied to synthesize a mass exchanger network (MEN) for minimizing phenol waste in an oil refinery. Phenolics are considered to be one of the major organic toxic species that should be minimized in waste streams in a refinery. Phenol-containing waste streams are generated from a number of processes in an oil refinery, such as a phenol solvent-extraction process and a catalytic cracking process *(15)*.

Problem Analysis. In the phenol solvent-extraction process, the waste stream leaving the process usually contains excessive phenol. These streams essentially come from three process units, i.e., raffinate tower, water/phenol tower, and extract stripper as illustrated in Figure 4. Conventionally, these streams are mixed first and then enter an absorber in which heated lubricating oil stocks absorb phenol from them *(15)*. In reality, the temperature, compositions, and flow rates of these streams are always different, and the fluctuations of these variables are stream independent. Consequently, mixing these streams is thermodynamically inefficient.

A mass balance computation indicates that lubricating oil alone is incapable of reducing the phenol concentration in the mixed stream to a tolerable range (≤ 0.002). This is especially true when severe disturbances appear at the inlet of the stream. Consequently, a MEN is desirable to reduce the concentration of phenol species regardless of the existence of various disturbances.

While a family of solvents can be used for recovering phenol, activated carbon is utilized in this case; however, the minimum quantity of activated carbon is expected to reduce the total cost. The process data for the synthesis problem is in Table I. Three streams from the raffinate tower, water/phenol tower, and extract stripper are designated rich streams R_1, R_2, and R_3, respectively; the lubricating oil and activated carbon are lean streams L_1 and L_2, respectively. Note that the mass flowrate and phenol concentration at the outlet of stream L_2 in the table are the upper limits. The intensity of disturbances appearing at the inlets of streams and the requirement of control precision of phenol concentrations at the outlets of streams are also specified. The target is to synthesize a cost-effective and highly controllable MEN. The pinch point is located at the lean end of the composite curve in a concentration-mass load of phenol species diagram. The minimum number of process units to be utilized is five. The minimum consumption of mass separating agent (MSA) is 0.0681 kg/s.

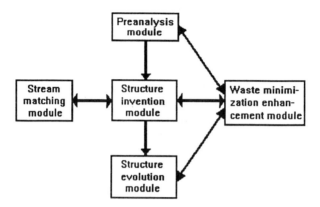

Figure 3. Connection mode of the modules in a synthesis procedure.

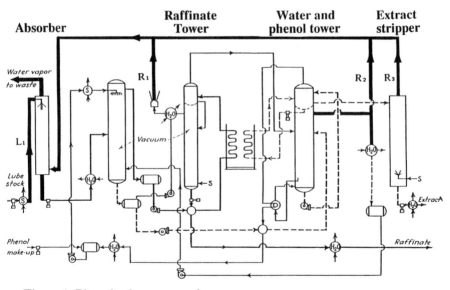

Figure 4. Phenol solvent-extraction process.

Table I. Specification of the MEN Synthesis Problem for Phenol Waste Minimization

Rich stream	Mass flowrate M (kg/s)	Source conc. y_p^s	Target conc. y_p^t	Max. deviation of mass flowrate of component p δM_p (kg/s)	Max. allowable deviation of target conc. δy_p^t
R1	4.3	0.15	0.002	0.025 (moderate disturbance)	0.0001 (high precision)
R2	1.5	0.10	0.001	0.012 (severe disturbance)	0.0001 (high precision)
R3	3.2	0.07	0.0015	0.020 (severe disturbance)	0.0003 (moderate precision)

Lean stream	Mass flowrate M (kg/s)	Source conc. x_p^s	Target conc. x_p^t	Max. deviation of mass flowrate of component p δM_p (kg/s)	Max. allowable deviation of target conc. δx_p^t
L1	12	0.002	0.08	0.026 (moderate disturbance)	0.01 (low precision)
L2	≤11	0.0007	≤0.01	0.001 (slight disturbance)	0.005 (low precision)

Solution and Comparison. By the present approach, an optimal solution of the synthesis problem, i.e., solution A, is identified in Figure 5. This MEN contains three absorbers and two adsorbers which are the minimum number of process units. The consumption of activated carbon also reaches the minimum. Moreover, it is highly controllable.

To demonstrate the superiority of the solution, two types of exhaustive search are conducted. For the first type of search, the same restrictions are imposed as those used in identifying a solution in Figure 5. Three solutions of this type are identified. One is the same as the solution structure in Figure 5, and the other two, named E-1-a and E-1-b, are depicted in Figures 6a and b, respectively. The second type of search is performed by imposing only one restriction, i.e., the minimum number of process units. In this case, the consumption of MSA is allowed to exceed the minimum requirement, if a process is highly controllable. With these restrictions, a large number of solutions are identified. For simplicity, only the two best, named E-2-a and E-2-b, are illustrated in Figures 7a and b, respectively.

The superiority of solution A in Figure 5 over the other four solutions E-1-a, E-1-b, E-2-a and E-2-b can be understood through examining their structures. In this synthesis problem, two intense disturbances exist at the inlets of rich stream R_2 and R_3. The concentration of phenol species at the outlets of rich stream R_1 and R_2 should be controlled very precisely, and that of R_3 should be controlled moderately. In solution A, these two intense disturbances at R_2 and R_3 cannot propagate to the outlets of R_1 and R_2, respectively; the intense disturbance at R_2 is not able to reach the outlet of R_3. Thus, the target concentrations (≤ 0.002) of phenol species in R_1, R_2, and R_3 can be effectively controlled. By contrast, the intense disturbance at the inlet of R_3 will propagate to the outlet of R_2 in structures E-1-a and E-2-a; the intense disturbance at the inlet of R_2 will reach the outlet of R_3 in structures E-1-b and E-2-b. Clearly, these four solutions are structurally undesirable.

The five solutions are compared in terms of the consumption of MSA, the number of process units, and WM capability which is reflected by the degree of structural controllability. As summarized in Table II, solution A is clearly the best one. The phenol concentrations of streams leaving the MEN can be strictly controlled to the minimum while yielding the minimum cost.

Concluding Remarks

Waste minimization and management are becoming the key issues today in chemical and petrochemical industries in complying with the regulations of environmental protection. To effectively minimize waste from sources in these industries, one of the most important ways is to evaluate the existing process structures and modify them if necessary, or design certain sub-processes. An artificial intelligence approach has been developed in this work for this purpose. By this approach, broad knowledge required for the design activity, whether it is symbolic or numeric, can effectively be acquired, represented and manipulated, thereby facilitating the design of a cost-effective process with minimal waste

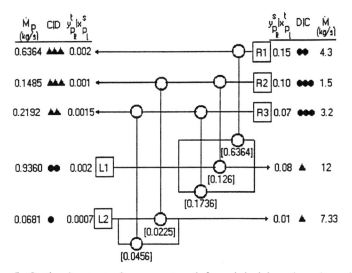

Figure 5. Optimal mass exchanger network for minimizing phenol species in waste streams (identified by the AI approach).

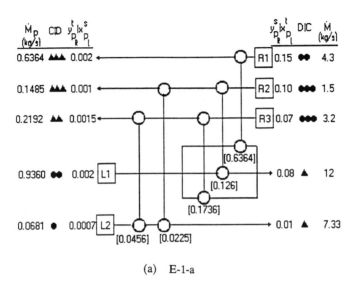

(a)   E-1-a

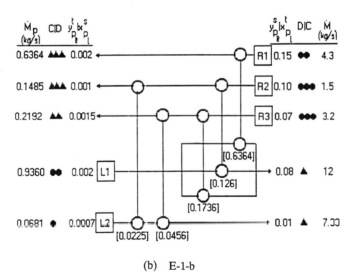

(b)   E-1-b

Figure 6. Mass exchanger networks for minimizing phenol species in waste streams (identified by the first type of exhaustive search).

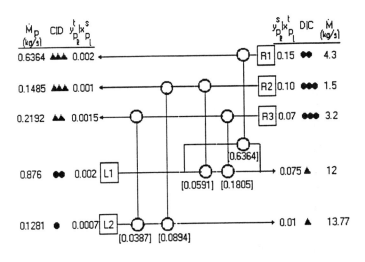

(a)   E-2-a

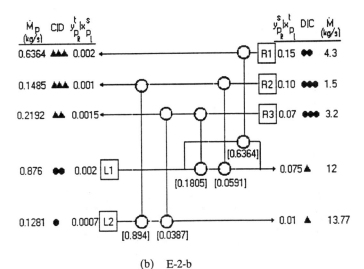

(b)   E-2-b

Figure 7. Mass exchanger networks for minimizing phenol species in waste streams (identified by the second type of exhaustive search).

Table II. Comparison of the Solutions of MEN's for Phenol Waste Minimization

Criterion	A (AI approach)	E-1-a	Solution E-1-b (exhaustive search)	E-2-a	E-2-b
Consumption of mass separating agent (kg/s)	0.0681	0.0681	0.0681	0.1281	0.1281
Number of mass tranfer units	5	5	5	5	5
WM capability (degree of structural controllability)	0.773	0.682	0.636	0.636	0.636

generation. The efficacy of the approach has been demonstrated by designing a mass exchanger network for minimizing phenol waste streams in an oil refinery. This approach is currently being applied to the modification of a process to minimize hydrogen sulfide species in waste streams, also in an oil refinery.

Literature Cited

1. Hazardous Waste Engineering Research Laboratory. *Waste Minimization Opportunity Assessment Manual,* U.S. EPA; Cincinnati, OH, 1988.
2. Higgins, T. *Hazardous Waste Minimization Handbook,* Lewis Publ., MI, 1989.
3. El-Halwagi, M. M; Manousiouthakis, V. *AIChE J.,* **1989**, 35, 1233-1244.
4. El-Halwagi, M. M; Manousiouthakis, V. *AIChE J.,* **1990**, 36, 1209-1219.
5. Berglund, R. L.; Lawson, C. T. *Chem. Eng.,* **1991**, 120-127.
6. Douglas, J. M. *Ind. Eng. Chem. Res.,* **1992**, 31, 238-243.
7. Huang, Y. L. *Integrated Process Design and Control of Chemical Processes via Artificial Intelligence Techniques: Application to Waste Minimization,* Ph.D. Thesis, Kansas State University, Manhattan, Kansas, 1992.
8. Barr, A.; Feigenbaum, E. A. *The Handbook of Artificial Intelligence;* HeurisTech Press: Stanford, CA, 1982; Vol. 1, pp. 3-11.
9. Zimmermann, H. J. *Fuzzy Set Theory - and Its Application,* Kluwer-Nijhoff: Hingham, MA, 1985.
10. Dey, H. M. *Hazardous Waste Minimization: Potential and Problems for Texas Refineries,* M.P.A. Report, University of Texas at Austin, TX, 1990.
11. Linnhoff, B.; Townsend, D. W.; Boland, D.; Hewitt, G. F.; Thomas, B. E. A.; Guy, A. R.; Marsland, R. H.; Flower, J. R.; Hill, J. C.; Turner, J. A.; Reay, D. A. *User Guide on Process Integration for the Efficient Use of Energy,* The Institute of Chemical Engineering, London, 1982.
12. Liu, Y. A. In *Recent Development in Chemical Process and Plant Design;* Editors, Y. A. Liu; H. A. McGee, Jr.; W. R. Epperly; Wiley: New York, NY, 1987, pp. 147-260.
13. Linnhoff, B.; Kotjabasakis, E. *Chem. Eng. Prog.,* **1986**, 82, 23-28.
14. Huang, Y. L.; Fan, L. T. *Comp. & Chem. Eng.,* **1992**, 16, 497-522.
15. Nelson, W. L. *Petroleum Refinery Engineering,* 4th Ed.; McGraw-Hill: New York, NY, 1969, pp. 360-362.

RECEIVED December 15, 1993

Chapter 7

Crystallization of Mechanical Pulp Mill Effluents through Hydrate Formation for the Recovery of Water

Cathrine Gaarder, Yee Tak Ngan, and Peter Englezos

Pulp and Paper Centre, Department of Chemical Engineering,
The University of British Columbia, Vancouver,
British Columbia V6T 1Z4, Canada

This paper presents data toward the development of a new crystallization technology for the concentration of pulp mill effluents and the subsequent recovery of water. In this process, clathrate hydrate crystal formation is the crystallization step. A screening of clathrate-hydrate forming substances was performed and two substances were chosen, carbon dioxide and propane. Experiments were carried out with these substances to measure the temperature and pressure conditions at which hydrates form in mechanical pulp mill effluent. It was determined that hydrate crystals can form in these effluents at temperatures several degrees above the normal freezing point of water. The measurements were compared with similar hydrate formation data in pure water. It was found that the presence of impurities in the effluent samples did not cause any appreciable change in the equilibrium hydrate formation conditions.

The pulp and paper industry requires a large amount of water to run its unit operations and, as a result, generates a substantial volume of dilute aqueous effluent that contains a variety of impurities of both organic and inorganic matter (*1*). In the past few years the discharge limits on these effluents have become increasingly stringent. And, as a consequence, the industry must improve the current effluent treatment technologies or develop new ones that can substantially reduce the pollution load that is now being discharged into the environment.

The current technologies that are being used include primary and secondary treatment systems. Primary systems, e.g. clarifying tanks, physically remove the majority of the settleable solids in the effluent whereas secondary systems, predominantly biological, remove a large portion of the organic compounds resulting in a decrease in the biological oxygen demand, BOD, of the effluent. The mills that have secondary end-of-pipe treatment systems in place are capable of complying with the current regulations but if these regulations become more stringent, as expected, these systems may prove to be inadequate (*2*).

0097–6156/94/0554–0114$08.00/0

Therefore, a two-fold approach to implement innovative water management strategies that would result in zero-liquid discharge (ZLD) mills is emerging. The first part is to modify the internal process to minimize the consumption of fresh feed water and the second is to develop cost effective end-of-pipe technologies that can recover clean water from the effluent. The separation technologies that are being considered for these ZLD systems include evaporation, membrane separation, and freeze concentration, also referred to as freeze crystallization (*2*). Compared with other separation technologies the major advantage of a crystallization process is that at a given temperature and pressure only one component will crystallize and it will do so as a pure crystal (*3*). The separation process that is the focus of this paper, clathrate hydrate concentration, is a variant of freeze concentration.

Freeze concentration refers to the concentration of an aqueous solution by generating a solid phase within the liquid stream, i.e. freezing it, followed by the physical separation and melting of the resulting crystals (*3*). This process is currently used by the food industry, e.g. the production of frozen orange juice, and in the desalination of seawater (*4*). The freeze process has the potential to be made more economical if the freezing step is modified to generate clathrate hydrates (clathrate hydrate concentration). The reason being that clathrate hydrates (*5*), also referred to as hydrates, have the ability to form at temperatures several degrees above the normal freezing point of water. Clathrate hydrate crystals are formed at suitable pressure and temperature conditions by the physical combination of water molecules with a large number of different molecules, e.g. argon, carbon dioxide and propane, which can be trapped within a network of hydrogen bonded water molecules (*6, 7*). The gas hydrate crystals contain only water and the hydrate forming substance. The concentration of aqueous streams by this process was patented by Glew (*8*) and was also investigated by Werezak (*9*). Formation of hydrate crystals in seawater was considered as the basis of a process to recover pure water. The process was demonstrated on a pilot plant stage (*10, 11, 12, 13*) but did not become commercial for economic reasons. The application of the clathrate process has recently become of interest in waste minimization as well as in the food products industry (*4, 14, 15*).

In this work, we are investigating the formation of clathrate hydrates in mechanical pulp mill effluents which are aqueous streams containing organic and inorganic substances. The scope of our work is to develop a separation technology based on clathrate-hydrate crystal formation. The process will concentrate the effluent and recover clean water to be recycled back into the mill. In the present study, our objectives are: (a) to select two suitable hydrate forming substances based on a number of criteria; (b) to form hydrates in mill effluents and observe the characteristics of the crystal formation; (c) to determine the minimum pressure, at a given temperature, at which hydrate crystals are formed; and (d) to discuss a conceptual process for the recovery of clean water from the mill effluent.

Background

Effluents from Mechanical Pulping. In mechanical pulping the wood chips are fed through parallel refiner discs which mechanically break down the wood into individual fibres. There are several different mechanical pulping processes. For example, pulp produced by the thermomechanical pulping (TMP) process is preheated and refined under pressure whereas bleached chemi-thermomechanical pulp (BCTMP) is produced by the addition of chemicals in the preheater, refined under pressure, and then bleached. The organic composition of the wood chips depends on their age and the wood furnish used. The general composition includes extractives (neutral solvent-soluble components), lignins (acid-insoluble aromatic compounds), hemicelluloses (mild acid or alkali-soluble carbohydrates), and cellulose (strong alkali insoluble carbohydrates). The effluent that results from these processes contains the wood material that was not recovered as fibre and traces of the chemicals that are added during the process. These components are either dissolved in the aqueous media or appear as small particles or colloidal material.

Clathrate Hydrate Phase Diagram. A partial phase equilibrium diagram for the propane-water system in the hydrate formation region is depicted in Figure 1. The points indicate experimental data from Kubota et al. (*16*) whereas the lines are interpolations of these data. Line KL is the vapor pressure line for propane. Line AQ defines the locus of hydrate-vapor (rich in propane)-liquid-w(rich in water) phase equilibrium. At a given temperature the pressure that corresponds to a point on this line is the minimum pressure required for hydrate formation. This pressure is often referred to as the incipient hydrate formation pressure. Propane hydrate is formed at conditions corresponding to pressures and temperatures on the left of line AQ. The temperature corresponding to point Q is the maximum temperature at which propane can form hydrate. This point is called the upper quadruple point. At this point four phases namely, hydrate, vapor, liquid-w and liquid-p(rich in propane) coexist in equilibrium. Line AQ extends down to another quadruple point (not shown on the plot) at 273.15 K where hydrate, ice, vapor and liquid water coexist in equilibrium. Line QC defines the locus of hydrate-liquid-w-liquid-p condensed phase equilibrium. Propane hydrate is formed when the pressure and temperature correspond to a point on the left of line QC. It should be noted that a similar diagram for carbon dioxide can also be prepared with data from the literature.

It is well known that the presence of impurities such as electrolytes in the water alters the location of the three phase line. In particular, at a given temperature, the required pressure for the formation of clathrate hydrate crystals is higher or at a given pressure, the equilibrium hydrate formation temperature is lower (depression of hydrate formation). For example, in Figure 1, the incipient hydrate formation conditions in a 2.5 wt % NaCl solution are shown. These experimental points are on the line A1Q1. The deviation from the formation conditions in pure water depends on the concentration of the impurities. Because the mill effluents contain various substances, our objective is to determine the locus of the hydrate-vapor-liquid equilibrium with the mill effluent as the liquid phase. This will be accomplished by measuring the pressure-temperature (P-T) hydrate formation conditions. Knowledge of these three phase equilibrium conditions is necessary for the design of the clathrate hydrate process.

Conceptual Clathrate Hydrate Concentration Process

There are at least four unit operations used to perform a separation with the clathrate hydrate concentration process. These are shown in Figure 2. They include the formation of the crystals, the separation of the crystals from the concentrate, the decomposition of the crystals, and the recovery of the hydrate former. Clathrate hydrate formation is accomplished by mixing the hydrate former with the effluent at suitable pressure and temperature conditions. The function of the separation step is to remove the clean hydrate crystals from the concentrated solution. This is a difficult separation (*10, 11*) because surface contaminants, impurities in the concentrated solution, are attracted to the crystal surface. The degree of this attraction will determine both the ease and extent of the separation. These attractions can be minimized if the surface area to volume ratio of the crystal is small. Once the hydrates are clean they are decomposed and the resulting mixture is separated into water and the hydrate former which is then recovered and returned to the hydrate formation unit.

Effluent Sources

The effluent samples used in this study came from two different mills. One sample was obtained from Quesnel River Pulp, in Quesnel, British Columbia, which produces both BCTMP and TMP, resulting in a mixed TMP/BCTMP generated effluent. The other sample came from Fletcher Challenge, in Elk Falls, British Columbia, which produces TMP. The chemicals used during the BCTMP runs at Quesnel include sodium sulphite, and DTPA (Diethylenetriaminepentaacetic Acid) which are used in the preheating stage, and caustic, sodium silicate, magnesium sulphate, DTPA, and a high consistency hydrogen peroxide which are used in the bleaching stage (Jackson, M., Sunds Defibrator, personal communication, 1992). The total solids content of the TMP effluent was found to be 710 mg/L, compared with 30,000 mg/L for the TMP/BCTMP effluent. The volatile solids portion increased from 240 mg/L to 12,000 mg/L for the TMP and TMP/BCTMP effluents respectively. These differences result from the specific processes and operations used at the two mills. It is well known that the use of chemicals in the BCTMP process will decrease the pulp yield, in comparison to the TMP process, resulting in an increase in the solids load of the effluent. The volume of water used in the process will also affect the solids concentration.

Selection of Hydrate Formers

Several criteria were used to determine which of the over one hundred hydrate formers would be most suitable for this study. The first criterion compared the hydrate formation equilibrium curves in pure water to determine which hydrate formers could form hydrates at temperatures well above 0°C and at pressures close to atmospheric. This was chosen as the initial criterion because the hydrate formation temperature must be well above the freezing point of the solution to give clathrate hydrate concentration an energy advantage over freeze concentration. The next set

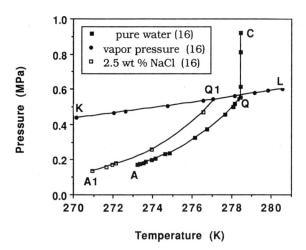

Figure 1. Propane-Water Partial Phase Diagram in the Hydrate Formation Region

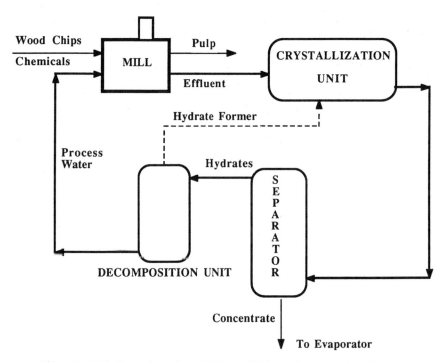

Figure 2. Unit Operations for a Clathrate Hydrate Concentration Process

of criteria were used to find environmentally acceptable hydrate formers, i.e. hydrate formers that are nontoxic, have low ozone depletion potential (ODP), low global warming potential (GWP), low explosion potential, and are nonflammable.

Other criteria that were considered include the solubility of the hydrate former in water, its cost and availability. The solubility aids in determining what type of a separation step, if any, is required to recover the hydrate former from the water phase once the hydrates have been decomposed. The extent of hydrate recovery will depend on both the cost of the hydrate former and the degree of process water purity that is required. Based on the above selection criteria the two hydrate formers chosen for this study were carbon dioxide and propane. Table 1 lists pertinent information for these compounds.

TABLE 1: Hydrate Former Characteristics for Propane and Carbon Dioxide

CHARACTERISTIC	CARBON DIOXIDE	PROPANE
Maximum equilibrium T & P in pure water (quadruple point)	10.1 °C at 4.5 MPa (*19*)	5.3 °C at 0.542 MPa (*16*)
Ozone Depletion Potential (ODP)	0	0
Global Warming Potential (GWP)	1.0	-
Explosion Potential	None	Lower level 2.1% Upper level 9.5%
Flammability	Nonflammable	Autoignition Temperature 432°C
Toxicity	Asphyxiant TLV 5000 ppm	Asphyxiant (no threshold level given)
Hydrate density (g/cm^3)	1.112	0.88
Solubility in water at maximum temperature (quadruple point)	6.1 wt percent	0.06 wt percent
Heat of hydrate formation (KJ/mol hydrate)	-55.0	-133.2
Number of water molecules per gas molecule in hydrate	7.3	17.95

Propane is known to be flammable and has an explosion potential if not handled properly, but it was decided that due to its wide spread use in other industries, procedures that would minimize both of these risks could be implemented. Carbon dioxide was chosen despite its high hydrate formation pressures, up to 45 atm, and its high solubility in water because it is inexpensive, nontoxic and does not pose a safety hazard. Other hydrate formers that were considered include cyclopropane and the new refrigerant R-134a. It was found that the cost of cyclopropane was much higher than that for propane, and that R-134a was both expensive and difficult to obtain in small quantities for lab purposes.

Experimental Apparatus and Procedure

The experimental apparatus that was used to study hydrate formation in mill effluents is described in detail elsewhere (17, 18). The main component of the apparatus is the vessel in which the hydrates form. Two different vessels were used. The low pressure vessel was used for propane hydrate formation whereas the high pressure one was used for carbon dioxide hydrate formation. In both cases, the vessel sits inside a glycol-water bath in which the temperature is controlled by an external refrigerator/heater, and a uniform temperature maintained by a motor driven stirring mechanism. Both vessels had sight windows such that the hydrate formation process could be viewed. The procedure used to determine the incipient hydrate phase equilibrium data was the same for both of the vessels. At standard procedure to obtain such data was used(17).

Results and Discussion

Experimental Hydrate Formation Conditions. The objectives of the experimental part of this study were to determine if hydrates could form in mechanical pulp mill effluents and ,if so, how would the impurities in the effluents affect the formation conditions. Initial experiments were performed to ensure that hydrates would form in mechanical pulp mill effluents by using propane or carbon dioxide. It was found that both propane and carbon dioxide could easily form hydrates in the TMP/BCTMP effluent. Subsequently, the incipient equilibrium pressure at temperatures above the normal freezing point of water were determined.

The results of the incipient equilibrium hydrate formation experiments determined for the TMP/BCTMP and the TMP effluent samples in the presence of propane are shown in Figure 3. Results for pure water are also shown to give a comparison. It was found that hydrates formed in both of the effluents at pressures only slightly above those needed for hydrate formation in pure water at the same temperatures. Figure 3 also shows that hydrates will form in both of the effluent samples at very similar conditions. This is a positive characteristic when considering the implementation of the clathrate hydrate concentration process as it implies that the process can accommodate changes in effluent composition.

The experiments using carbon dioxide as the hydrate former showed similar results to those with propane. The results are shown in Figure 4. Again the TMP/BCTMP effluent formed hydrates at pressures only slightly above those for pure water at the same temperatures. Also found in Figure 4 are results from a run in which pure water was used. The pressure-temperature diagram that resulted from

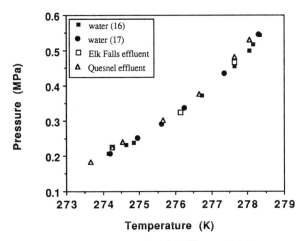

Figure 3. Hydrate Formation in Effluent with Propane.

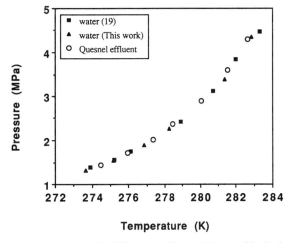

Figure 4. Hydrate Formation in Effluent and Pure Water with Carbon Dioxide.

these experiments were in good agreement with data from the literature thereby validating the experimental equipment and procedure used in this study.

The results from both hydrate formers have shown that the impurities in the mill effluent did not significantly affect the hydrate formation conditions when compared with those in water. This implies that either their concentration is too low to cause any appreciable change in the hydrate formation conditions or the effluent contains impurities that are not hydrate inhibitors. If indeed hydrate inhibitors exist in the effluent then it is expected that as water is removed and becomes a part of the hydrate phase the concentration of the impurities will increase. This will result in an increase in the pressure needed for hydrate formation to occur. The extent that this will affect the formation conditions depends on the amount of water that is to be recovered by the process. For example, recovery of 80 percent of the water will increase the concentration of the impurities 5 times. In this case, one expects the inhibiting action would be roughly five times that in the initial effluent. However, because the inhibiting action of the impurities in the initial effluent was small the overall increase due to the concentration process should again be small.

Qualitative observations about the crystals were made in experiments where considerable amount of hydrates were formed. The crystals that were formed in the presence of propane varied in colour, the ones that floated on top looked to be clear, similar to ice, compared to those at the bottom and on the sides of the vessel. The reason for the colour of these crystals is not known for sure but it may be due to occlusion of concentrate in the vacant space between individual crystals. These observations point out the need for more research into the formation process (kinetics and morphology studies) before a large scale crystallization unit can be designed.

Conclusions

Propane and carbon dioxide were selected as suitable hydrate forming substances for the clathrate hydrate formation process. Propane was found to form hydrates in both types of mechanical effluents that were examined. Carbon dioxide was tested with only one of the effluents and was found to form hydrates readily. The pressure-temperature hydrate formation conditions (partial phase diagram) for both gases were determined. It was found that the presence of the impurities in the mill effluents altered the pressure-temperature equilibrium locus only slightly.

Acknowledgments

The financial support from the Natural Science and Engineering Research Council of Canada (NSERC) and the Pulp and Paper Research Institute of Canada (PAPRICAN) is greatly appreciated.

Literature Cited

1. Smook, G.A. *Handbook for Pulp and Paper Technologists*; 2nd Edition; Angus Wilde: Vancouver, B.C., 1992 pp 45-64, 378-391.

2. Beaudoin, L.; Dessureault, S.; Barbe, M.C. *CPPA 77th Annual Meeting*, Montreal, Quebec, 1991, A235-A253.

3. Heist, J.A. *Chem. Eng.* **1979**, May, pp 72-82.

4. Douglas, J. *EPRI Journal.* **1989**, Jan- Feb. pp 16-21.

5. Findlay, R.A. In *New Chemical Engineering Separation Techniques*; Schoen, H.M., Ed.; Interscience Publishers: New York, 1962; pp. 257-318.

6. Englezos, P.; Kalogerakis, N.E.; Bishnoi, P.R. *J. Inclusion Phenom. Mol. Recognit. Chem.* **1990**, 8, pp 89-101.

7. Sloan, E.D. *Clathrate Hydrates of Natural Gases*; Marcell Dekker: New York, 1990, pp 1-66.

8. Glew, D.N. Solution Treatment, U.S. Patent 3,085,832, Oct.16, 1962.

9. Werezak, G.N. *AIChE Symp. Ser.* **1969**, 65(91), pp 6-18.

10. Knox, W.G.; Hess, M.; Jones, G.E.; Smith, H.B. *Chem. Eng. Prog.* **1961**, 57(2), pp 66-71.

11. Barduhn, A.J. *Desalination*, **1968**, 5, pp 173-184.

12. Barduhn, A.J. *Chem. Eng. Prog.* **1975**, 71(11), pp 80-87.

13. Tleilmat, B.W. In *Principles of Desalination*; Spriengler, K.S., Ed.; 2nd Edition; Academic Press: New York, 1980, pp 359-400.

14. *Tech Commentary* **1988**, 1(1), pp 1-4 (Freeze Concentration).

15. *U.S.D.Energy Report* **1988**, Prepared by Heist Engineering. Report available from NTIS, U.S.D.Commerce, Springfield, Virginia 22161 (Beet Sugar Refining Applications).

16. Kubota,H.; Shimizi, K.; Tanaka, Y.; Makita, T. *J. Chem. Eng. Japan.* **1984**, 17, pp 423-428.

17. Englezos,P.; Ngan, Y.T. *J. Chem. Eng. Data.* **1993**, 38, pp 250-253.

18. Englezos,P.; Ngan, Y.T. *Fluid Phase Equilib.* **1993**, in press.

19. Robinson,D.B.; Mehta, B.R. *J. Can. Petr. Tech.* **1971**, 10, pp 33-35.

RECEIVED December 1, 1993

Chapter 8

Extraction of Mercury from Wastewater Using Microemulsion Liquid Membranes
Kinetics of Extraction

Karen A. Larson and John M. Wiencek

Department of Chemical and Biochemical Engineering, Rutgers, The State University of New Jersey, P.O. Box 909, Piscataway, NJ 08855-0909

A microemulsion containing a cation exchanger reduces the mercury content of an aqueous phase from 500 ppm to 0.3 ppm. This is a 40-fold improvement over equilibrium extraction. Extraction kinetics are first order in mercury concentration and zero order with respect to oleic acid concentration and pH which is consistent with film theory predictions for an instantaneous reaction that is mass transfer controlled. A model of the separation has been formulated and includes mercuric ion mass transfer to the droplet surface, equilibrium between aqueous mercury and organic mercury complex at the droplet interface, diffusion of the complex through the organic phase, and stripping of mercury at the internal droplets. Without the use of adjustable parameters, this model predicts mercury extraction rate and equilibrium.

Since the discovery of the toxicity of mercury to humans, the effects of mercury contamination in aquatic environments and sediments has been the subject of major scientific investigations (*1*). In addition to contaminated water, there is growing concern over leaching of heavy metals, especially mercury, from landfills to nearby groundwater. Discarded batteries are a major source of mercury in landfills. In fact, household batteries account for 88% of all mercury found in municipal solid waste, or about 1.4 million pounds (*2*). The possibility exists that some landfills could be a potential source of mercury in water. As a result, there is a need to not only rectify aquatic regions that are known to be contaminated with mercury, but also to have the technology available to treat aqueous streams that may become contaminated at some future time.

Several methods exist for recovery of mercury from aqueous streams: precipitation of mercury from a caustic solution of mercuric sulfide in the presence of aluminum (*3*); precipitation of mercuric oxide from caustic solutions (*4*); and solvent extraction of mercury ions using ion exchangers (*5*). With all of these processes, the remediation of one contaminated stream results in the formation of another, requiring either landfilling or additional aqueous stream treatment (e.g. aluminum precipitation of mercury). Treatment such as electroplating is preferred since the recovered metal can be reused; however, the concentration of mercury in some aqueous waste streams is too low for efficient use of this technique.

0097–6156/94/0554–0124$08.00/0

Trace contaminants can be efficiently extracted from waste streams using emulsion liquid membranes, a separation technique that combines extraction and stripping into a single step thereby minimizing equilibrium limitations inherent with conventional solvent extraction (*6*). When an ion exchanger is incorporated into the formulation, coarse emulsion liquid membranes have been used to extract a number of metals from aqueous solutions: silver, gold, and palladium ions using macrocyclic crown ethers as the organic phase carrier (*7*), extraction of zinc and chromium ions using Aliquat 336 as a carrier (*8*), extraction of copper using LIX reagents (*9-11*), and extraction of mercury ions using either dibutylbenzoylthiourea (*12*) or oleic/linoleic acid (*13*) as carriers. Recently, Larson and Wiencek (*14*) have utilized microemulsions to conduct an extraction of mercury from water.

A microemulsion is an optically transparent, thermodynamically stabilized dispersion of two immiscible phases. Such systems offer potential advantages when used as a liquid membranes over a coarse emulsion liquid membranes which are only kinetically stable (*15*). A microemulsion forms spontaneously when the components (typically consisting of an aqueous phase, organic phase, surfactants, and other additives) are brought into contact. While the mechanism of extraction is similar to that of coarse emulsion separations, the physical nature of the emulsion phase is quite different. For example, the internal droplet size in a coarse emulsion is on the order of 1 mm while the droplet size range in a microemulsion is 50-1000 Å. The interfacial tension between the organic and aqueous phases of a microemulsion is generally less than 1 dyne/cm; an order of magnitude less than that measured for coarse emulsions (*16*). The smaller internal droplet size and lower interfacial tensions ultimately result in much faster mass transfer rates due to increased surface area for mass transfer and reaction. The thermodynamic stability of the microemulsion can result in reduced leakage and coalescence for some formulations, and finally, microemulsions can be emulsified or demulsified by changing the temperature, for example. The disadvantage is that the aqueous content of the microemulsion is limited by thermodynamic constraints. For example, a microemulsion containing an internal phase consisting of 1N NaOH does not leak when used as a liquid membrane; however, if 10 N NaOH is desired, a microemulsion will not form at all. In contrast, a coarse emulsion can be made with either internal phase. Microemulsions have been used to separate acetic acid from water (*15*) and copper ions from water using a microemulsion containing benzoylacetone (*17*).

In order to successfully model the extraction process, both the equilibrium and kinetics of the metal:liquid ion exchanger reaction must be determined and related to the other transport processes occurring during a liquid membrane separation (mass transfer, diffusion, reaction). The liquid ion exchanger chosen for this work is oleic acid because it forms a microemulsion which is stable when used as a liquid membrane (*14*). In this paper, the extraction and stripping of mercury and oleic acid is characterized as a function of mercury, oleic acid, modifier concentration and pH. Equilibrium models are then developed by identifying the appropriate aqueous and organic phase reactions. The kinetics of the mercury:oleic acid reaction are then evaluated as a function of mixing, mercury, oleic acid and hydrogen ion concentration and compared to film theory models for two phase reactions. Finally, a diffusion/reaction model is developed for extraction of mercury with oleic acid in a batch stirred tank reactor.

Experimental

Materials. The chemicals used in this study were: oleic acid (food grade, Fisher), mercuric nitrate monohydrate (Fisher), tetradecane (Fisher and Humphrey Chemical) and sulfuric acid (Fisher). Surfactants and modifiers were: Igepal CO-210 (Rhone-Poulenc), Igepal CO-430 (Rhone-Poulenc), DM-430 (GAF), and decanol (Kodak).

Procedures. Experiments to determine the distribution of mercury salts between aqueous and organic phases were carried out in test tubes. Unless otherwise noted, equal volumes of organic and aqueous phases were mixed on a tube rotator for a minimum of 1 hour at ambient temperature (18-23°C) and then centrifuged on a laboratory centrifuge for five minutes to disengage the phases. The mercury concentration in the aqueous phase was measured using the mercury hydride procedure on a Perkin-Elmer Model 3030 Atomic Absorption Spectrophotometer at a wavelength of 253.7 nm. The concentration of the organic mercury salt was calculated by material balance.

Kinetics of mercury extraction with oleic acid were evaluated in a modified Lewis cell shown in Figure 1. In these experiments, 175 ml of aqueous solution was added to the cell. The organic phase, consisting of tetradecane, oleic acid and 10 w/w% CO-210 surfactant, was poured on top of the aqueous phase. Great care was taken to minimize the mixing of the two phases. The agitation was started (Lightnin DS1010 motor) and 2 ml samples of the aqueous phase were taken every minute for ten minutes. In order to maintain a constant liquid level in the cell, two ml of DI water was added to the aqueous phase via the sampling port after each sample was taken. The samples were analyzed for mercury by atomic absorption.

For the purpose of modeling the kinetics and measuring drop size, the model experimental system of Skelland and Lee (18) was used. The stirred tank reactor consisted of an 8 liter glass tank having a diameter of 22.5 cm and a height of 25 cm and containing four aluminum baffles having dimensions of 2x22 cm. Mixing was accomplished using a single stage, six flat blade Rushton turbine powered by a Lightnin Model DS1010 motor. Photographs of the dispersed oil phase were taken using an Olympus Boroscope Model OES F100-OHH-000-30 attached to an Olympus Model OMPC 35 mm camera. The light source was an Olympus Model KMI-5 Multifunctional Light Source fitted with a fiber optic cable. A schematic representation of the tank set-up is shown in Figure 2.

Microemulsions were formulated by equilibrating tetradecane containing 0.35 M oleic acid, stripping reagent (sulfuric acid) with a surfactant, DNP-8, an ethoxylated dinonylphenol having an average of 7 ethylene oxide groups per molecule. Equilibration time was 4-12 hours. The aqueous content of the microemulsion was analyzed using a Mettler Model DL18 Karl Fisher titrator.

Results and Discussion

Microemulsion Liquid Membrane Extraction of Hg^{+2}. To demonstrate the efficiency of microemulsion liquid membrane extraction compared to conventional solvent extraction, a microemulsion was formulated with 0.35 M oleic acid in tetradecane, 10 w/w% DNP-8, and 6 N sulfuric acid. At equilibrium, a clear microemulsion phase and excess aqueous phase could be observed. The aqueous content of the microemulsion phase was 11 w/w %. Typical separation kinetics are shown in Figure 3. The organic phase in the control experiment consisted of 0.35 M oleic acid in tetradecane with 10 w/w% CO-210 with no internal phase. Extraction of the feed phase with this formulation reduced the mercury content to 8.2 ppm . However, the feed phase after extraction with the microemulsion liquid membrane was reduced to 0.23 ppm. These results illustrate a major advantage of liquid membrane separations: more efficient extractions in a single stage because of the favorable shift in equilibrium resulting from the use of an internal phase that simultaneously strips mercury from the organic phase while it is being extracted from the bulk aqueous phase. Since the eventual goal is to model this kinetic behavior, it is necessary to first characterize the extraction and kinetics of the mercury:oleic acid system (without the internal phase). The rest of the discussion will encompass this characterization.

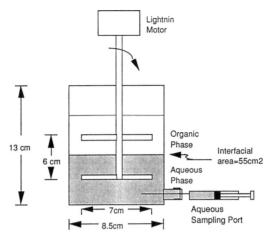

Figure 1. Schematic Representation of the Stirred Cell for Kinetic Studies

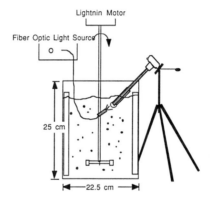

Figure 2. Schematic Representation of the Batch Reactor and Photographic Setup

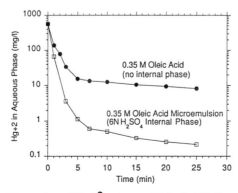

Figure 3. Extraction of Hg^{+2} using an Oleic Acid Microemulsion

Equilibrium Extraction and Modeling (No Internal Phase). When halide anions are not present in the aqueous phase, mercury exists primarily as Hg^{+2}. The extraction of mercury into tetradecane containing 0.10 M and 0.35 M oleic acid as a function of aqueous phase equilibrium pH is illustrated in Figure 4. The shape of the curves is characteristic of metal extraction behavior with cation exchangers (*19*). At low pH, very little mercury is extracted. As the pH is increased, extraction efficiency improves markedly. The maximum extraction occurs at an equilibrium pH of about 2.5. Extractions at pH higher than about 2.8 are not possible because of the formation of the insoluble oxide salt of mercury. As expected, the higher concentration of oleic acid shifts the equilibrium curve towards lower pH values. Similar trends were observed for Hg^{+2} extractions using capric acid and di(2-ethylhexyl) phosphoric acid (D2EHPA) (data not presented here). The extraction stoichiometry is determined performing loading experiments at high pH (around 2.5) where mercury extraction with oleic acid extraction is most efficient. Extraction levels off at a 1:4 ratio of mercury to oleic acid.

Modeling the extraction behavior requires identification of the aqueous phase reactions as well as consideration of the observed reaction stoichiometry in the organic phase. Since mercuric nitrate was used as the source of mercury and pH adjustments were accomplished with nitric acid and sodium hydroxide, the aqueous phase reaction equilibria considered were:

$$Hg^{+2} + NO_3^- \quad <--> \quad HgNO_3^+ \qquad logK=0.33 \qquad (1)$$
$$Hg^{+2} + 2NO_3^- \quad <--> \quad Hg(NO_3)_2 \qquad logK=-1.36 \qquad (2)$$
$$Na^+ + H_2O \quad <--> \quad NaOH + H^+ \qquad logK=-14.20 \qquad (3)$$
$$Na^+ + NO_3^- \quad <--> \quad NaNO_3 \qquad K=1.0 \qquad (4)$$

The value for the equilibrium constants are from the literature (*20*). Aqueous phase activity coefficients were estimated using the modified form of the Debye-Huckel equation:

$$Ln\ \gamma = -0.509*Z^2 \left[\frac{I^{0.5}}{(1.0 + I^{0.5})} \right] + 0.15\ I \qquad (5)$$

where Z is the charge of the ion and I is the solution ionic strength (*20*). Determination of the organic phase reaction with oleic acid was difficult since very little characterization of oleic acid under these conditions has been reported. In general, carboxylic acids form dimers in non-polar organic liquids as a result of intermolecular hydrogen bonding (*21*). By taking into consideration dimer formation of the oleic acid as well as the experimentally observed stoichiometry of 1:4 (mercury:oleic acid), the following form of the organic phase reaction is suggested:

$$Hg^{+2} + 2(HR)_2 \quad <--> \quad Hg(R \cdot HR)_2 + 2H^+ \qquad (6)$$

These equations, together with material balances for mercury, sodium, nitrate, oleic acid and charge, were solved numerically. The equilibrium constant for the organic phase reaction which gave the best fit of the data was 0.45+/-0.13. Model calculations are shown as solid lines in Figure 4.

Surfactant and Modifier Effects. While the extraction efficiency of the oleic acid/tetradecane system is quite high and the aqueous solubility of the organic phase is low, one disadvantage of this particular system is the formation of a solid phase, especially at high mercury loadings. This solubility problem is common when aliphatic solvents are used and can be overcome with the use of a modifier such as a long chain alcohol or nonylphenol (*21*).   Several equilibrium experiments were performed to determine the effect of modifiers (decanol, CO-210 and CO-430) in the oleic acid/tetradecane system on extraction equilibrium as well as solid phase formation. CO-210 and CO-430 are ethoxylated nonylphenols having 1.5 and 4 ethylene oxide groups per molecule, respectively.   These surfactants were chosen for two reasons: they have a low water solubility; consequently, they do not form emulsions during an extraction;   and, they are similar in structure to the surfactant required for the stabilization of the microemulsion system described earlier.

The effect of the presence of 10 %wt CO-210 in a solution containing 0.32 M oleic acid in tetradecane on mercury extraction equilibrium is evident in Figure 4.   The equilibrium curve has shifted markedly to the left; thus, drastically reducing the pH dependency of the extraction as well as eliminating solid phase formation.   A 10% solution of CO-210 in tetradecane (no oleic acid) did not extract mercury to any significant extent.   The effects of CO-210, CO-430 and decanol on mercury extraction are shown in Figures 5-a and 5-b.   Because of the high extraction efficiency of oleic acid, the effects of the addition of these modifiers on extraction is most readily seen by looking at the residual mercury content of the aqueous phase.   Consequently, the y-axis data are plotted as a ratio of mercury concentration in the aqueous phase without surfactant to that with surfactant.   This ratio represents an extraction enhancement effect (a ratio greater than 1 implies improved extraction as a result of the presence of the modifier).   Modifier concentration is plotted on the ordinate.   Figure 5-a shows that all three modifiers, when added with oleic acid, improve mercury extraction from the aqueous phase.   The shape of the curves are similar.   When the data are plotted as a function of oxygen content of the modifier, as shown in Figure 5-b, all three curves collapse onto a single curve suggesting that the enhancement effect is due in some way to the synergistic effects of the polar oxygen groups.   In fact, the general shape of the curve can be modeled by postulating the formation of a dimer of the mercury/oleic acid complex of the form:

$$2\,HgR_2 \cdot 2RH\ +\ Oxygen \longleftrightarrow (HgR_2 \cdot 2RH)_2 \tag{7}$$

In this case, the hydrophilic groups on the modifiers play a role in facilitating the formation of the dimer.   Formation of a complex dimer in the organic phase is consistent with the observed behavior for copper and nickel extraction using aliphatic carboxylic acids such as capric acid (*22-24*).

Mercury Stripping from Oleic Acid.   Mercury can be efficiently back-extracted from oleic acid using a strong mineral acid such as nitric, hydrochloric, or sulfuric acid.   Mercury stripping from oleic acid using 6N sulfuric acid at an organic:aqueous ratio of 10 (the same concentration as that used in the microemulsion formulation) is shown in Figure 6.   The scatter in the data is attributed to the high concentrations of mercury used in the study since small errors in the analysis of the aqueous phase could result in larger errors in the organic phase calculation.   At the higher concentrations, some precipitation was also observed.   The equilibrium modeling of the back-extraction takes into account the following aqueous phase reactions:

$$H_2SO_4 \longleftrightarrow H^+ + HSO_4^- \qquad LogK=1.98 \tag{8}$$
$$HSO_4^- \longleftrightarrow H^+ + SO_4^{-2} \qquad LogK=-1.98 \tag{9}$$
$$Hg^{+2} + SO_4^{-2} \longleftrightarrow HgSO_4 \qquad LogK=1.41 \tag{10}$$
$$H_2O \longleftrightarrow H^+ + OH^- \qquad LogK=-14 \tag{11}$$

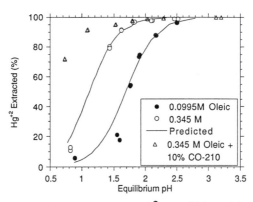

Figure 4. Extraction of Hg^{+2} with Oleic Acid

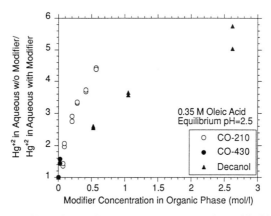

Figure 5-a. Effect of Modifiers on Mercury Extraction with Oleic Acid

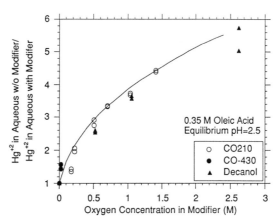

Figure 5-b. Effect of Oxygen Content in the Modifier on Extraction of Mercury with Oleic Acid

The equilibrium constants were taken from the literature (*20*). Since the organic phase included 0.35 M oleic acid as well as 10 w/w% CO-210 surfactant, the organic phase reactions include the complex dimer reaction (equation 7) with LogK=3.2 as well as:

$$HgR_2 \cdot 2RH + 2H^+ <--> Hg^{+2} + 2\ (HR)_2 \qquad (12)$$

The equilibrium constant for stripping which best fit the data was 32.8+/-7.1.

Mercury Extraction Kinetics. Initial experiments for determination of mercury:oleic acid extraction kinetics were carried out in the 8 L stirred tank reactor using a 10:1 aqueous:organic ratio; however, the extraction rate was so fast that >90% extraction was complete in less than 1 minute. Consequently, experiments were conducted in the modified Lewis cell which maintains a considerably smaller interfacial area between the two phases than a stirred tank, thus slowing down the reaction kinetics.

Figure 7 illustrates the effect of mixing speed on extraction rate. Extraction rate increases with mixing speed over the range 50-100 rpm. Above 100 rpm, some turbulence at the aqueous/organic interface was observed. Below 40 rpm, the agitation could not be maintained at a constant rate. The effect of initial mercury concentration in the aqueous phase on the initial extraction rate is shown in Figure 8. Here, the rate of extraction has been calculated as follows:

$$R_o = [Hg]_o \frac{dx}{dt} \frac{V_{aq}}{A} \frac{1}{60} \frac{1}{1000} \quad [mol/cm^2/s] \qquad (13)$$

where V is the volume of the aqueous phase and A is the interfacial area. The plot has a slope of 1.2 which is interpreted to be first order dependence with respect to mercury concentration over the concentration range 0.001-0.004 mol/l. In Figure 9, the extraction rate is plotted as a function of pH at several different mixing speeds. As in Figure 7, the overall rate of extraction at each pH increases with mixing speed, however, at a given mixing speed, the rate of extraction is essentially independent of pH over the range 1.80-2.55. In addition, all the lines are parallel indicating that there is no transition between different regimes or mechanism. In addition, Figure 10 shows the initial extraction rate as a function of oleic acid concentration. Here again, the extraction rate is essentially independent of oleic acid concentration over the range 0.035-0.35 M. All of the above concentration ranges are within the range that will be employed in the microemulsion formulations.

These kinetics can be interpreted using film theory predictions for two phase reactions (*25*). In fact, film theory was used to interpret copper extraction with benzoylacetone in a Lewis cell (*26*). In attempting to determine which of the reaction regimes is controlling, several assumptions are made: (1) The reaction is irreversible; (2) Hg^{+2} is insoluble in the organic phase; (3) oleic acid is sparingly soluble in the aqueous phase and as a consequence, the reaction is homogeneous; and, (4) there is no mass transfer resistance in the organic phase. Assumption 1 is reasonable considering that only initial rate data was used to calculate the extraction rate and the data over the 10 minute time interval is linear. The experimental conditions overwhelmingly favor the forward extraction (mercury into the organic phase). Assumption 2 is valid because control experiments show that Hg^{+2} is not extracted into tetradecane without the presence of oleic acid. The solubility of oleic acid in water was estimated by measuring the total organic carbon of an aqueous sample after prolonged contact with the organic phase. The solubility is approximately 1×10^{-4} mol/l which is considerably lower than typical 'sparingly soluble' materials (1×10^{-2} mol/l) (*27*). This would imply that the reaction probably occurs at the interface rather than in the bulk aqueous phase. The assumption of no mass transfer in the organic phase is valid because of the excess of

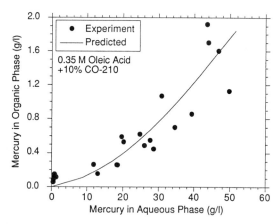

Figure 6. Mercury Stripping from Oleic Acid using 6N Sulfuric Acid

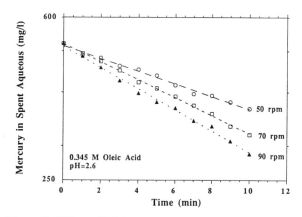

Figure 7. Effect of Mixing on Mercury Extraction Kinetics

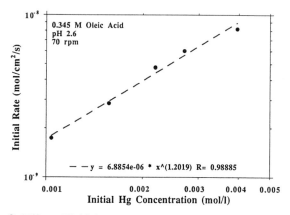

Figure 8. Effect of Initial Mercury Concentration on Rate of Extraction

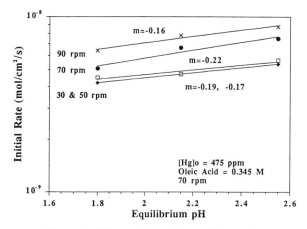

Figure 9. Effect of pH on Rate of Extraction

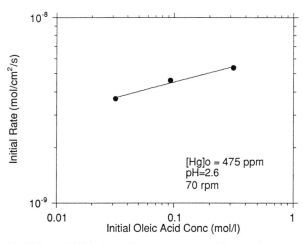

Figure 10. Effect of Oleic Acid Concentration on Mercury Extraction Rate

oleic acid compared to that of mercury as well as the favorable distribution coefficient for the mercury:oleic acid complex.

Film theory predicts four possible reaction regimes depending upon the relative rates of diffusion and reaction. (1) *Very slow reaction*. In this regime, the reaction is so slow that diffusional factors are unimportant, only the kinetics. In this case, one would not expect to observe an effect of mixing on reaction rate as is the case for the mercury:oleic acid system. (2) *Slow Reaction*. If our system were in this regime, the concentration of the oleic acid in the bulk aqueous phase would be zero, but would be finite in the mass transfer film. The rate expression in this case would be first order in oleic acid concentration. In our system, however, the rate is independent of oleic acid concentration. (3) *Fast Reaction*. The oleic acid would be depleted within the mass transfer film, but Hg^{+2} would not. This regime is a special case in that a first order reaction in oleic acid would have an observed rate that would be proportional to oleic acid concentration. If the reaction were zero order in oleic acid, the observed rate would be proportional to the one half power of oleic acid. Again, since our observed rate is independent of oleic acid concentration, this is most likely not the reaction regime. (4) *Instantaneous Reaction*. The reaction is so fast that it is assumed to occur in an infinitesimally thin plane. In this case, the rate would be independent of oleic acid concentration but would be proportional to Hg^{+2} concentration and the aqueous phase mass transfer coefficient. This is consistent with the first order behavior of mercury concentration shown in Figure 8. In fact, given this interpretation of the data, the mass transfer coefficients for the Lewis cell can be calculated from the slope of the data at each agitation rate in Figure 7. The mass transfer coefficients range from 0.001 to 0.004 cm/s which is within the range of mass transfer coefficients determined for Lewis cells (*28-29*).

Modeling Mercury Extraction Kinetics in a Stirred Tank Batch Reactor. Based on the information developed in the previous sections, the following assumptions are made: 1) Mercury extraction with oleic acid is a mass transfer limited process. 2) At the aqueous/organic interface, mercury is in equilibrium with oleic acid following equation 6. 3) As mercury diffuses into the drop, the complex dimerizes according to equation 7. The mathematical description of this process is as follows:

Bulk Phase

$$-V_e \frac{\partial(Hg_e)}{\partial t} = \frac{3}{R} (V_m + V_i) D \frac{\partial C}{\partial r}\bigg|_{r=R} \tag{14}$$

$$= V_e k_L a (Hg_e - Hg_s)$$

where:

$$a = \frac{3}{R} \left(\frac{V_i + V_m}{V_e} \right) \tag{15}$$

The following boundary and initial conditions apply:

at t=0, $Hg_e = Hg_0$

for $r \leq R$, C=0, $B = B_0$

$$Hg_s = \frac{C_s[H^+]^2}{K_{eq}[HR_2]^2} * \frac{\gamma_{H^+}{}^2}{\gamma_{Hg}} \tag{16}$$

<u>Organic Phase:</u> In the organic phase, the transport of free oleic acid is modeled by:

$$V_m \frac{\partial B}{\partial t} = V_m D_B \left[\frac{\partial^2 B}{\partial r^2} + \frac{2}{r} \frac{\partial B}{\partial r} \right] \tag{17}$$

The transport of the mercury:oleic acid complex is modeled by:

$$V_m \frac{\partial C}{\partial t} = V_m D_C \left[\frac{\partial^2 C}{\partial r^2} + \frac{2}{r} \frac{\partial C}{\partial r} \right] - 2V_m \frac{\partial C_2}{\partial t} \tag{18}$$

C_2 represents the complex dimer that is formed in the organic phase as a result of the presence of oxygen in the surfactant according to the following reaction:

$$2 C + Oxy = C_2 \qquad\qquad K_{eq} = 1600 \ (l/mol)^2 \tag{19}$$

or:

$$C_2 = K_{eq}[Oxy] [C]^2 \tag{20}$$

This expression can now be substituted into equation (18) . The initial conditions for these equations are:

at t=0, $B=B_0$, C=0, and $C_2=0$ for the globule interior.

Because the oxygen concentration of the 10% wt. CO-210 is approximately 0.7 mol/l compared to the typical mercury complex concentration in the organic phase of 10^{-5} mol/l, the oxygen concentration can be taken to be constant over the course of the extraction. Consequently, a new equilibrium constant can be defined as follows:

$$K' = K_{eq}[Oxy] \qquad (l/mol) \tag{21}$$

therefore:

$$\frac{\partial C_2}{\partial t} = K' \frac{\partial}{\partial t}[C^2] = 2K' C \frac{\partial C}{\partial t} \tag{22}$$

The boundary conditions are:

$$\text{at } r=0, \qquad \frac{\partial B}{\partial r} = \frac{\partial C}{\partial r} = 0 \tag{23}$$

$$\text{at } r=R, \qquad k_L \left(Hg_e - Hg_s \right) = -D_B \frac{1}{2} \frac{\partial B}{\partial r} = D_c \frac{\partial C}{\partial r} \tag{24}$$

The following dimensionless variables are defined:

length scale $X = r/R$
time $t = D_c \, t/R^2$
bulk Hg $A = Hg/Hg_o$
oleic acid $BB = B/B_o$
complex#1 $CHG = 2C/B_o$
complex#2 $C2 = 4C2/B_o$

volume fraction $Phi = \dfrac{V_i + V_m}{V_e}$

Biot number $Bi = \dfrac{k_L R}{D_c}$

The dimensionless form of the equations become:

External Phase

$$\frac{\partial A}{\partial t} = -3 * Phi * Bi \, (A - A_s) \tag{25}$$

at t=0, A=1

$$A_s = \frac{C_s \, H^2}{K_{eq} \, (B_o - 2C_s)^2} \frac{\gamma_H^2}{\gamma_{Hg}} \tag{26}$$

Globules

$$\frac{\partial BB}{\partial t} = r_{BC} \left[\frac{\partial^2 BB}{\partial X^2} + \frac{2}{X} \frac{\partial BB}{\partial X} \right] \tag{27}$$

$$[1 + 2K'B_o CHG] \frac{\partial CHG}{\partial t} = D_C \left[\frac{\partial^2 CHG}{\partial X^2} + \frac{2}{X} \frac{\partial CHG}{\partial X} \right] \tag{28}$$

at t=0, BB=1, CHG=0, C2=0

$$\text{at } X=0, \qquad \frac{\partial BB}{\partial X} = \frac{\partial CHG}{\partial X} = 0$$

$$\text{at } X=1, \qquad \frac{\partial CHG}{\partial X} = \frac{2 \, Hg_{o,e} Bi}{B_o} \, (A - A_s)$$

The model equations were solved using an implicit finite difference method. Two simplifying assumptions have been made. {1} The pH of the bulk is set at the equilibrium pH. It is not a changing variable. This assumption is reasonable because the pH range at which mercury extraction from the bulk occurs is in the range that is most efficient for mercury extraction (it is in the horizontal portion of the %Hg Extracted vs. pH curve of Figure 4). In other words, it is in the range where mercury extraction is not very sensitive to pH. Fixing the equilibrium pH will not affect the overall rate of extraction in the model since the initial extraction rate is severely mass transfer limited. {2} The diffusivity of the free oleic acid and the complex are assumed to be equal. The diffusivity for the oleic acid:mercury complex was calculated using the Wilke-Chang (*30*) correlation and was found to be 1.00×10^{-6}. The droplet size in the tank was measured photographically and the Sauter mean diameter was calculated to be 0.0076 cm at 350 rpm. The aqueous phase mass transfer coefficient was calculated using the Skelland and Lee (*18*) correlation for low interfacial tension solutions and was found to be 0.00331 cm/s. It is important to note that this value is within the range of mass transfer coefficients in the Lewis cell experiments where the mercury extraction rate was concluded to be mass transfer limited.

Results of the model calculations are shown in Figures 11 and 12. Figures 11-a and 11-b show the effect of Biot number on mercury extraction from the bulk solution at short and long times. As expected, lower values of the Biot number (corresponding to a lower value for the mass transfer coefficient) result in slower extraction kinetics. The curve corresponding to a Biot number of 25.1 has been calculated for our system using the measured value for the Sauter mean diameter and correlations for the diffusivity and mass transfer coefficient described previously. This value results in a predicted extraction rate that agrees quite well with the experimentally determined extraction rate (points). At longer times, the model predicts a final mercury concentration of 15.8 ppm while the experimentally determined equilibrium concentration is 20 ppm. Considering that there are no adjustable parameters in this model, and that all parameters that have been used were either measured by independent experiments or determined from separate correlations, agreement between theory and experiment is good. Figures 12-a, 12-b, and 12-c illustrate the effect of Biot number on the transport process occurring within the organic phase. Specifically, the complex dimer concentration is monitored as a function of time and Biot number. For example, with a Biot number of 7.58, transport of mercury into the drop is so slow that the diffusion process for the dimer maintains nearly a constant concentration throughout the drop. External mass transfer is significant for the entire extraction process. As the Biot number increases, dimer concentration at the surface also builds up. This is especially evident at Biot number of 75.8. Over time, however, as equilibrium in the bulk phase is approached, the extraction rate slows down and, as a result of diffusion and equilibrium within the drop, a nearly constant dimer concentration is attained throughout the drop.

Conclusions

Equilibrium extraction and stripping of mercury with oleic acid has been characterized as a function of pH, mercury, oleic acid and modifier concentration. The equilibria have been modeled by assuming a mercury:oleic acid complex of the form: $HgR_2 \cdot 2RH$ having an extraction equilibrium constant of 0.449. Extraction of mercury is enhanced by the presence of surfactant or modifier. The effect has been postulated to be complex dimer formation in the organic phase. While this behavior is consistent with that observed for other metals, spectroscopic studies are needed to confirm this structure. Kinetics of mercury extraction were found to be first order in $[Hg^{+2}]$, and zero order in [pH] and [Oleic acid]. This behavior is consistent with instantaneous

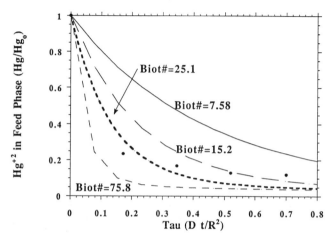

Figure 11-a. Mercury Extraction in a Batch Stirred Tank:
Theory vs. Experiment for Short Time

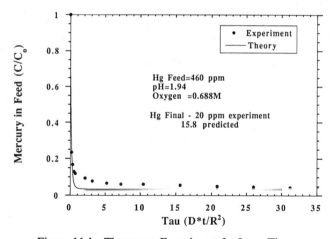

Figure 11-b. Theory vs. Experiment for Long Times

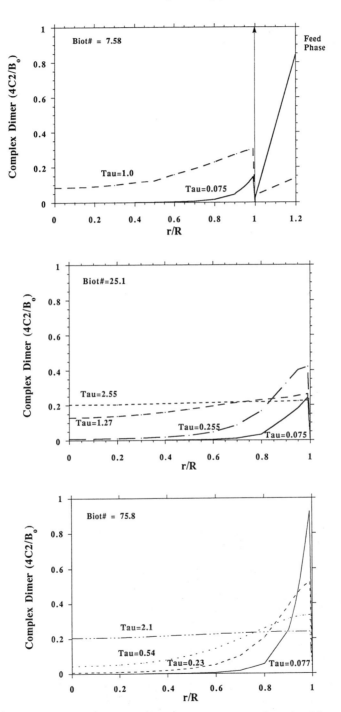

Figure 12. Effect of Biot Number on Distribution of Complex Dimer
Within Drop

reaction kinetics that are mass transfer limited. A diffusion/ reaction model for the extraction kinetics which accounts for these observations has been developed for a batch stirred tank system. Using independently measured or calculated values for various parameters, this model predicts mercury extraction kinetics and equilibrium quite well. Future work will be directed at incorporating the internal phase reaction into the model.

Nomenclature

B   oleic acid concentration- $(HR)_2$ (mol/cm^3)
C   mercury/oleic acid complex #1 $(HgR_2 \cdot 2RH)$ (mol/cm^3)
C2   mercury /oleic acid dimer (mol/cm^3)
D   diffusivity (cm^2/s)
γH hydrogen ion activity coefficient
γ_{Hg} mercury ion activity coefficient
Hg   mercury ion (mol/cm^3)
K_{eq} equilibrium constant
k_L mass transfer coefficient (cm/s)
Oxy   oxygen concentration in surfactant (mol/cm^3)
R   radius (cm)
t   time (s)
V   volume (cm^3)

Subscripts

e   external or bulk phase
i   interface or surface
m   membrane phase (organic phase)
o   initial concentration
s   surface

Acknowledgments. This work has been sponsored by the USGS Cooperative Program in Water Resources, Hazardous Substance Management Research Center of New Jersey (NSF/Industry Cooperative Research Center) and Merck & Company, Inc.

Literature Cited

1. Jaffar, M; Ashraf, M.; Pervaiz, S., *Pakistan J. Sci. Ind. Res.,* **1988** *31,* 322.
2. Holusha, J., "Keeping a Gadget-Mad Nation Charged Up - and Safe," *New York Times,* **1991**.
3. Butler, J.N., *Nevada Bureau of Mines, Report#5,* **1963**.
4. Peters, R. W.; B. Mo Kim (eds.), Separation of Heavy Metals and Other Trace Contaminants, AIChE Symposium Series, no. 243, v. 81, **1985**, p.165.
5. Rousseau, R., Handbook of Separations Processes, Chapter 8, **1987**, p. 467.
6. Li, N.N., *US Patent* 3.410.794, 1968.
7. Izatt, R.; Clark, G.; Christensen, J., *Sep. Sci. Tech.,* **1987**, *22,* 691.
8. Fuller, E.J.; Li, N.N., *J. of Mem. Sci,* **1984**, *22,* 251.
9. Kondo, K.; Kita, K., *J. Chem. Eng. Japan,* **1981**, *12,* 20.
10. Volkel, W.; Halwachs, W; Schugerl, K., *J. of Mem. Sci .,* **1980**, *6,* 19.
11. Achwal, S.; Edrees, S., *Arabian J. Sci. Eng.,* **1988**, *13,* 331.
12. Weiss, S.; Grigoriev, V., *J. of Mem. Sci.,* **1982**, *12,* 119.
13. Boyadzhiev, L.; Bezenshek, E., *J. of Mem. Sci.,* **1983**, *14,* 13.

14. Larson, K.; Wiencek, J., *I&EC Res.* **1991**, *31*, 2714.
15. Wiencek, J.M.; Qutubuddin, S., *Colloids and Surfaces*, **1989**, *29*, 119.
16. Wiencek, J.M.; Qutubuddin, S., *J. of Mem. Sci.*, **1989**, *45*, 311.
17. Wiencek, J.M.; Qutubuddin, S., *Sep. Sci. Tech.*, **1992**, *27*, 1407.
18. Skelland, A.; Lee, J., *AIChEJ*, **1981**, *27*, 99.
19. Chapman, T., "Extraction-Metals Processing, " Handbook of Separation Process Technology, R. Rousseau (ed) John Wiley and Sons, New York, **1987** p. 467.
20. Lindsay, W., Chemical Equilibria in Soils, John Wiley and Sons, New York, **1979**.
21. Ritcey, G.; Ashbrook, A. Solvent Extraction: Principles and Applications to Process Metallurgy, Part I,, Elesevier, New York, **1984** p. 26,
22. Tanaka, M.; et. al., *J. Inorg. Nucl. Chem.*, **1969**, *31*, 2591.
23. Tanaka, M.; Niinomi, T., *J. Inorg. Nucl, Chem.*, **1965**, *27*, 431.
24. Kojima, I.; et. al., *J. Inorg. Nucl. Chem.* **1970**, *32*, 1333.
25. Sharma, M. "Extraction with Reaction," Handbook of Solvent Extraction, M. Baird and C. Hanson (eds.), Wiley Interscience, New York **1983** p 37.
26. Kondo, K.; Takahashi, S.; Tsuneyuki, T; Nakashio, F, *J. Chem. Eng. Jap.*, **1978**, *11*, 193.
27. Nanda, A.; Sharma, M., *Chem. Eng. Sci.*, **1966**, *21*, 707.
28. Nanda, A.; Sharma, M., *Chem. Eng. Sci.*, **1967**, *22*, 769.
29. Danckwerts, P., Gas-Liqud Reactions, McGraw-Hill, New York, **1970** p 181.
30. Reid, R.; Prausnitz, J.; Sherwood, T., The Properties of Gases and Liquids, McGraw-Hill, New York, **1977**.

RECEIVED December 3, 1993

Chapter 9

Removal of Gasoline Vapors from Air Streams by Biofiltration

W. A. Apel[1], W. D. Kant[2], F. S. Colwell[1], B. Singleton[2], B. D. Lee[1], G. F. Andrews[1], A. M. Espinosa[1], and E. G. Johnson[1]

[1]Center for Bioprocessing Technology and Environmental Assessment, The Idaho National Engineering Laboratory, Idaho Falls, ID 83415–2203
[2]Industrial Division, EG&G Rotron, Saugerties, NY 12477

Research was performed to develop a biofilter for the biodegradation of gasoline vapors. The overall goal of this effort was to provide information necessary for the design, construction and operation of a commercial gasoline vapor biofilter. Experimental results indicated relatively high amounts of gasoline vapor adsorption can occur during initial exposure of the biofilter bed medium to gasoline vapors. Biological removal occurred over a 22 to 40°C temperature range with removal being completely inhibited at 54°C. The addition of fertilizer to the bed medium did not increase rates of gasoline removal in short term experiments in the relatively fresh bed medium used. Microbiological analyses indicated that high levels of gasoline degrading microbes were naturally present in the bed medium and that additional inoculation with hydrocarbon degrading cultures did not appreciably increase gasoline removal rates. At lower gasoline concentrations the vapor removal rates were considerably lower than those at higher gasoline concentrations indicating that substrate availability (i.e. transport) was limiting in the system. This implies that system design facilitating gasoline transport to the microorganisms could substantially increase gasoline removal rates at lower gasoline vapor concentrations. Preliminary results from tests of a field scale prototype biofiltration system showed volumetric productivity (i.e. average rate of gasoline degradation per unit bed volume) values consistent with those obtained with the columns at similar inlet gasoline concentrations. In addition, total BTEX removal over the operating conditions employed was 50-55%. Removal of benzene was approximately 10-15% while removal of the other members of the BTEX group was much higher, typically > 80%.

The use of gas/vapor phase-grown microorganisms to remove organic components from gas streams is a technology that is currently under active development. Initial information indicates that gas/vapor phase bioreactors may be suitable for the selective removal of a number of different gases and vapors from gas streams, and as such, have a major role to play in the efficacious, cost-effective remediation of volatile pollutants from a number of different environmental media.   This view is supported by the work of researchers in the United States together with the results from field applications of gas and vapor bioprocessing in Europe, particularly in Germany and The Netherlands (*1, 2, 3, 4, 5, 6, 7, 8*). Most of these European applications have employed a particular type of gas/vapor phase bioreactor known as a biofilter.

In concept, biofilter design is relatively simple.   The biofilter consists of a filter bed within a physical container, often times a box or a column which may be either closed or open (Figure 1).   A gas distribution system is present to route the gas/vapor stream through the bed.   The bed usually consists of a soil or soil/compost mixture which supports the growth of microorganisms capable of metabolizing the gas or vapor to be treated.   During operation, the bed is maintained in a moist, but non-saturated condition, to enhance growth of the microbes which form a thin biofilm on the surface of the bed medium, while allowing maximum gas transport to the bed.   The gas or vapor stream to be treated is passed through the bed where the microbes present in the biofilm metabolize the gas or vapor.   The end products of this metabolism are typically carbon dioxide, water, additional microbial biomass, and, depending on the composition of the gas stream treated, inorganic salts.

As many as 500 biofilter applications are believed to be active in Europe (*8*). Many of these applications are directed towards odor control from a variety of sources including agriculture, food processing, slaughter houses, sewage treatment and rendering plants.   Other applications include control of volatile toxic compounds from chemical plants, coating operations, and foundries.

Biofiltration ideally complements vacuum vapor extraction (VVE) technology which is becoming widely accepted for the remediation of volatile contaminants from both vadose and saturated soil zones (*9*).   VVE operations result in a vapor-rich off-gas stream that often requires further treatment before release into the atmosphere.   This treatment typically consists of filtration of the off-gas stream through activated carbon or incineration (*10, 11*).   While effective, both of these treatments are expensive and may constitute up to 50% of the cost of an ongoing VVE operation.   Biofiltration is an alternative VVE off-gas treatment (Figure 2) since numerous VVE off-gas components like hydrocarbons and halocarbons can be attacked and degraded microbiologically.

Hydrocarbon degradation by both pure cultures and consortia of microorganisms is well documented.   Under aerobic conditions, microbes can mineralize to CO_2 essentially all hydrocarbon volatiles including aliphatics and aromatics following typical Michaelis-Menton reaction kinetics (*12, 13, 14*).   The goal of the research reported below was to generate the information necessary to design, construct and operate a prototype commercial biofilter using these organisms to remove gasoline vapors from VVE off-gas streams.

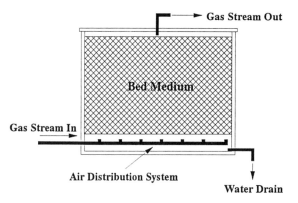

Figure 1. Schematic of closed biofilter.

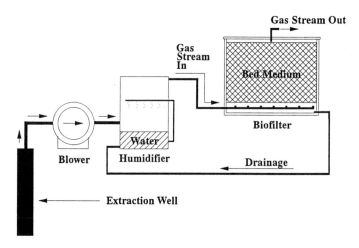

Figure 2. Schematic of vapor vacuum extraction off-gas treatment with a biofilter.

Materials and Methods

Bed Medium. The bed medium used for serum vial and column experiments was a proprietary blend which was developed to minimize gas flow channeling while facilitating the growth of microorganisms. In all experiments, sufficient water was added to the bed medium blend to create a soil suction of 10 centibars as measured with a 2100F Soilmoisture probe (Soilmoisture Equipment Corp, Santa Barbara, CA).

Terminology. In convention with commercial VVE terminology, gasoline vapor concentrations are reported in terms of parts per million (ppm). Unless otherwise specified, ppm concentrations were determined based on volume to volume comparisons.

Optimum Temperature. Four temperatures were screened for maximum gasoline vapor removal rates; 22, 30, 40 and 54°C. Five g of bed medium was added to 120 ml serum bottles and the bottles were sealed with 20 mm teflon lined rubber stoppers secured with aluminum crimp rings. Gasoline was added to the serum bottle headspace at a concentration of 400 ppm. Headspace samples of 1.0 ml were analyzed via gas chromatography (GC) for total hydrocarbon content over time. Likewise, samples of 250 μl were analyzed via GC to monitor changes in O_2 and CO_2 concentrations in the headspace. This and all other serum bottle experiments reported below were conducted in triplicate and the results were averaged, unless otherwise stated.

Inocula Testing. Two commercially available hydrocarbon degrading microbial cultures and a bed medium leachate were tested for enhancement of hydrocarbon removal in the biofilter medium. One commercial culture, Solmar L-103, was obtained from the Solmar Corporation, while another commercial culture, Tesoro PES-31, was obtained from the Tesoro Petroleum Distributing Company. The leachate was obtained from percolation of water through the bed medium. This was accomplished by filling a closed 3" X 36" glass column with bed medium and recirculating a gasoline-in-air mixture through the column in an upflow direction while simultaneously circulating a water drip through the column in a downflow direction. Gasoline and oxygen levels in the recirculating gas mixture were monitored with time via GC as described below, and a fresh gasoline-in-air mixture was reinjected into the column whenever either the gasoline or oxygen vapor concentration approached depletion. At the end of two weeks, leachate from the recirculating water drip was collected and used as a gasoline enrichment inoculum in subsequent experiments. All inocula were added to the bed medium as liquid cultures. Before addition, cultures were adjusted to a density of approximately 1.0×10^9 cells ml^{-1} and 0.26 ml of inoculum were added per g of bed medium. The two commercial inocula were exposed to gasoline vapors for approximately 10 days before addition to the bed medium. All inocula were tested for gasoline vapor removal versus uninoculated controls using serum vial cultures as described above and incubated statically at 30°C. Gasoline levels in the headspace of the serum vials were followed with time via GC analyses.

Nutrient Addition. A 0.5% solution of Miracle Gro fertilizer was added to the bed medium mixture to determine whether the addition of macro-nutrients (*e.g.* nitrogen, phosphorous, etc.) and trace elements had an effect on hydrocarbon degradation. The Miracle Gro was added to the bed medium in liquid form as recommended by the manufacturer to assure even distribution of the nutrients. The Miracle Gro-supplemented bed medium was then placed in serum vials as described above. Gasoline was added to the headspace and the samples were incubated at 30 °C. Samples were analyzed via GC for hydrocarbon remaining in the headspace.

Adsorption Isotherm. The ability of the bed medium to sorb gasoline hydrocarbons abiotically was determined at 30 and 40°C. At 40°C testing was done in a concentration range of 50 to 2,000 ppm gasoline-in-air. A concentration range from 400 to 25,000 ppm gasoline-in-air was tested at 30°C. The gasoline concentration in the headspace of samples at 30°C was monitored via GC analyses at 30, 60 and 90 min to determine when adsorption was complete as denoted by a decrease in the rate of hydrocarbon loss. The gasoline concentration in the headspace of the vials incubated at 40°C was analyzed via GC after incubation for approximately 3 h. Indigenous microbes in the bed medium used at both temperatures had been killed by the addition of 500 μg of $HgCl_2$ per g of bed medium. Adsorption isotherms were generated from the resulting data using previously described standard techniques (*15*).

Serum Bottle Kinetics Testing. After an optimal temperature range was determined kinetic experiments were conducted in serum vials using the techniques described above. Gasoline removal was monitored in a concentration range from 400 to 25,000 ppm gasoline-in-air. The experiments were run at a temperature of 30°C and analyzed via GC for concentration of gasoline hydrocarbons in the headspace at 48 h intervals.

Column Studies. Continuous flow studies were done using 3.5" by 3' glass columns containing 1.5 kg of bed medium. The columns were operated in a downflow mode at a flow rate of 1,000 ml min^{-1}. Gasoline vapor was added to the columns by sparging air through liquid gasoline, which was then diluted in air using a gas proportioner. The liquid gasoline stock was regularly replaced to maintain a constant feed of the more volatile components. The resulting gasoline vapor-in-air mixture was then fed directly through the columns. The columns were operated in a concentration range from approximately 100 to 15,000 ppm gasoline-in-air. Moisture addition was achieved by manual addition of water to the vapor stream. The columns were sampled using an automatic sampling system consisting of 2 - 16 valve actuators connected to a 3-way valve. The sample lines were 1/16" fused silica-lined stainless steel tubing. The samples were injected into the GC using a 6-way gas sampling valve equipped with a 1.0 ml sample loop. Samples were drawn by application of a vacuum to the sample lines.

Prototype Bioreactor. Field studies with a full scale prototype biofilter were conducted at a VVE installation located at a full service gasoline station site which had been in operation since 1928 (Figure 3). Subsurface soils and groundwaters were contaminated at this site with high gasoline levels being noted in samples taken from monitoring wells. The biofilter consisted of a skid mounted, cylindrical steel vessel, 1.52 m in diameter by 2.13 m in height (Figure 4). Five stainless steel plenums supporting 2.8 m^3 of bed medium were spaced at even intervals throughout the vessel. A gas stream consisting of gasoline vapors in air was removed from the VVE well by a blower system (EG&G Rotron, Saugerties, NY) and passed through a humidifier/flow conditioner (EG&G Rotron, Saugerties, NY) prior to passage in an upflow direction through the biofilter. Bed temperature was maintained between 25 and 35°C. The gas stream flow rate was 20 cfm. Inlet gasoline concentration varied from 2 - 2000 ppm (total petroleum hydrocarbons) while inlet relative humidity varied from 3-100%. Gasoline levels in the inlet and outlet gas streams were monitored as described below.

Analytical Methods for Serum Bottle and Column Studies. Headspace gasoline concentrations of the vials were determined by gas chromatographic analysis using a Hewlett-Packard 5890 Series II gas chromatograph (GC). For data collection, the GC was connected to a Compaq Deskpro 486 computer running Hewlett Packard 3365 Chemstation software. One ml samples of the vial headspace gases were injected into the GC which was equipped with a Restek 30m, 0.32 mm i.d. Rtx-5 column containing a 0.25 μm film thickness of crossbonded 95% dimethyl - 5% diphenyl polysiloxane. Following sample injection, the GC was maintained for 6 min at 30°C after which the temperature was increased to 150°C at a rate of 10°C min^{-1}. The injector temperature was 225°C and the flame ionization detector (FID) temperature was 275°C. Helium was used as the carrier gas at a flow rate of 1.6 ml min^{-1} at a split ratio of 1:8.

Oxygen and carbon dioxide in the headspaces were analyzed using a Gow Mac Series 550P GC equipped with a thermal conductivity detector and an Alltech CTRI column. Helium at a flow rate of 60 ml min^{-1} was the carrier gas, the injector and column temperatures were 30°C.

Analytical Methods for Prototype Field Biofilter Studies. Gasoline concentrations in the inlet and outlet gas stream of the prototype biofilter were analyzed chromatographically using an EG&G Chandler Engineering gas chromatograph equipped with an FID. The GC was connected to a Spectrophysics SP4400 integrator. Column temperature was 80°C while inlet and detector temperatures were both 85°C. The column used was a 10 ft 5% SP1200 and 1.75% Bentone 34 on Chromosorb W-AW 80/100. The carrier gas was N$_2$ at a flow rate of 30 ml min^{-1}. The unit was calibrated using Scott standards for BTEX and n-alkanes (C$_1$ - C$_7$).

Site plan showing layout of remedial system

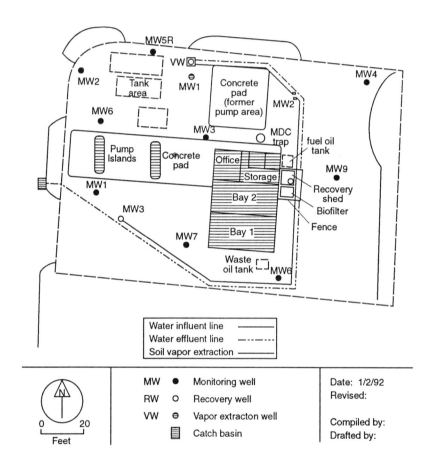

Figure 3. Schematic of gasoline station remediation site.

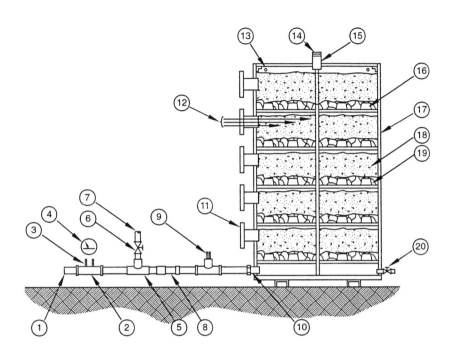

1. Contaminated air from system
2. Venturi flow meter
3. Pressure taps
4. Magnehelic gauge
5. Educter
6. Metering valve
7. H₂O in
8. Site tube
9. Relative humidity & temp. probes
10. Inlet
11. 4 in. ASTM flanges for instrumentation

12. Temperature, tensiometer, and pressure/sampling probes
13. Moisturizing loop, one per level (not used in normal operation)
14. Treated air out
15. Outlet
16. Stainless steel redistribution plenums
17. Carbon steel vessel 60 in. ID 0.25 thk epoxy coated
18. Biofilter medium
19. pine bark
20. Drain valve

Probe wiring omitted for drawing clarity

Figure 4. Schematic of prototype field biofilter.

Results and Discussion

The adsorption isotherms generated at 30 and 40° C with the sterile bed medium show gasoline adsorption as a function of gasoline concentration in the headspace at equilibrium.   Linear sorption patterns were noted over the gasoline concentrations tested. At 30° C the isotherm is described by the equation of $y = 1.02 \times 10^{-5} x$ with an $R^2 = 0.98$ (Figure 5), and at 40°C it is described by the equation $y = 1.19 \times 10^{-5}x$ with an $R^2 = 0.99$ (Figure 6). These data indicate that in the absence of microbial activity, the bed medium has a appreciable gasoline sorptive capacity and by itself can remove significant amounts of gasoline vapors from gasoline/air mixtures. Over the gasoline concentration range tested, gasoline sorption increased as a function of gasoline concentration in the headspace with approximately 60 μg gasoline being adsorbed per g of bed medium at a 5 g (m^3 of air)$^{-1}$ gasoline concentration in the headspace at 40°C. The sorption rates appeared to be comparable at both temperatures, and as such, would not be expected to vary significantly over the temperature range employed in these experiments or that anticipated with the field biofilters to be tested.

Gasoline vapor removal as depicted in Figure 7 further illustrates the sorptive capacity of the bed medium. Over the gasoline concentrations tested, an initial sharp decrease in headspace gasoline levels was observed. It is hypothesized that this initial decrease was due primarily to adsorption working in conjunction with biodegradation, after which a more gradual, prolonged decrease in headspace gasoline levels was noted. This gradual decrease was believed to primarily correspond to biodegradation of the gasoline by microbes indigenous to the bed medium, as corresponding gasoline loss in control vials was negligible. Studies with fresh bed medium showed $> 10^6$ colony forming units per g capable of growing in a gasoline atmosphere on agar mineral salts plates. As such, this bed medium appeared to have an adequate indigenous gasoline degrading microbial population such that an exogenous inoculum was not required for gasoline degradation.

Studies with fertilizer supplimented bed medium revealed no detectable increase in microbial activity. These studies were performed with fresh bed medium, and further studies will be conducted to determine whether fertilizer supplementation has a positive influence on microbial removal rates of aged bed medium which may be depleted of some essential nutrients.

Experiments conducted with bed medium inoculated with two commercially available blends of hydrocarbon degrading microbes and a third INEL-generated blend of hydrocarbon degrading microbes confirmed this hypothesis. The bed medium inoculated with these three exogenous inocula showed no increase in degradation rates versus uninoculated bed medium containing only indigenous microbes. This confirms that no inoculation of hydrocarbon biofilters is necessary upon startup due to the substantial levels of hydrocarbon degrading bacteria naturally contained in the bed medium. In a commercial field bioremediation application, this is a significant positive point since the culturing and addition of a stable, exogenous, gasoline degrading inoculum could be a significant complication in the start up of a biofilter and add appreciable expense and uncertainty to the operation.

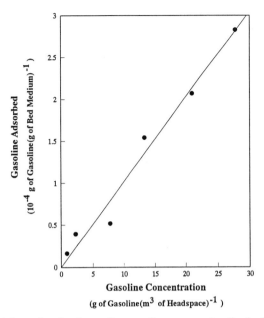

Figure 5. Adsorption isotherm for gasoline vapors by the bed medium at 30°C.

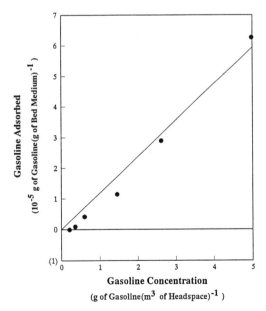

Figure 6. Adsorption isotherm for gasoline vapors by bed medium at 40°C

Figure 8 supports this view and furthermore, illustrates that within approximately 2 weeks, the bed medium can become further acclimated to gasoline vapors with a significant increase in gasoline degradation rates occurring after this acclimation period. After exposure for two weeks to 400 ppm gasoline in air, respiking of the same bed medium-containing vials with similar gasoline-in-air concentrations, results in significantly increased (i.e. 4X) gasoline vapor removal rates. Subsequent respikes do not lead to appreciably more rapid gasoline removal rates, and as a result, the bed medium appears to be fully conditioned to provide the most rapid rates of gasoline removal achievable after the initial two week conditioning period. This effect is believed to be the direct result of an increased population of gasoline degrading microorganisms being present in the bed medium after the 2 week initial exposure period and has been demonstrated previously in soil microbial populations that become adapted to organic vapors (16).

These data indicate that during startup of a biofilter under field conditions, a significant amount of gasoline removal should be noted initially both from adsorption and biodegradation, with removal rates potentially increasing for a period of days to weeks following startup. As a result, it may be possible to establish a relatively low gasoline feed rate during the startup/conditioning phase and after conditioning, to increase this feed rate 4-5X without any loss in gasoline removal efficacy. This concept needs to be further tested in the field-scale biofilter to ascertain how rapidly the conditioning occurs in a full sized biofilter operated under field conditions.

Figure 9 shows gas chromatograms illustrating removal and degradation of various gasoline components by biofiltration using the bed medium in serum vials. In general, with the analytical method employed (see Materials and Methods section), higher molecular weight fractions of the gasoline vapor appear at longer retention times on the gas chromatograms shown. In the serum vial batch systems, it is clear that biofiltration tends to remove the higher molecular weight constituents more quickly than the lower molecular weight constituents. The BTEX (i.e. benzene, toluene, ethylbenzene, and xylene) components were removed in a relatively rapid fashion with benzene being more recalcitrant to removal than the other members of this group. Based on the physical properties of the bed medium, it is reasonable to hypothesize that this differential removal of gasoline components is a property of the sorptive nature of the blend and subsequent degradation by the microorganisms as opposed to a physiological inability of the microbes to metabolize the lower molecular weight components. This suggests that to rapidly remove the lower molecular weight, more volatile gasoline components, either additional exposure time or a modified bed medium with greater sorptive properties will be necessary.

Temperature studies showed that microorganisms indigenous to the bed medium actively degrade gasoline vapors over a 22 to 40°C temperature range (Table I). There was very little difference in observed gasoline degradation rates over this range, however no degradation was observed at 54°C. From a practical standpoint, these data are encouraging when considering the operation of a gasoline vapor biofilter in the field since the 22 to 40°C temperature range more than encompasses operating temperature range anticipated under field conditions.

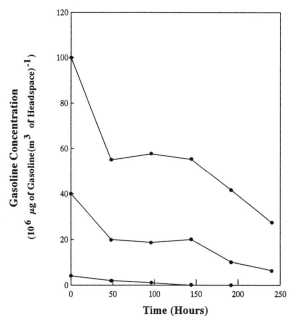

Figure 7. Removal of gasoline at various concentrations by unacclimated bed medium.

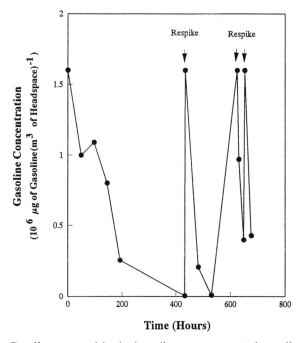

Figure 8. Gasoline removal by bed medium upon repeated gasoline exposure at 30°C.

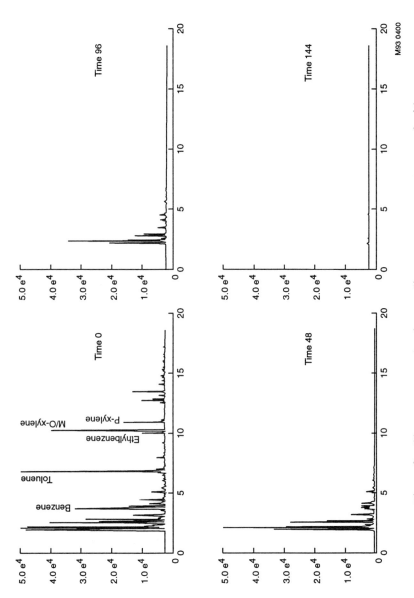

Figure 9. Chromatograms showing gasoline vapor component removal with time by the bed medium in serum vials at 30°C.

As such, the biofilter is expected to be robust relative to anticipated temperature fluctuations, and as a result, operating temperature should not be a factor requiring rigid control during field operation.

Table I. Influence of temperature on gasoline removal rates by microorganisms indigenous to the bed medium

Temperature (°C)	Gasoline Removal Rate (μg of Gasoline$(m^3$ of Compost)$^{-1}$ hr^{-1})
22	7.1 x 10^3
30	7.6 x 10^3
40	6.3 x 10^3
54	None Detected

When considering biofilter sizing for field applications it is pertinent to calculate the volumetric productivity of the biofilter at the substrate concentrations that will be encountered under actual operating conditions. This information in turn can be used to size the biofilter as a function of substrate feed rate. The volumetric productivity values obtained in serum vial experiments over a range of gasoline concentrations encompassing those expected to be encountered in actual field applications are shown in Figure 10. Similar data for the columns run at a flow rate of 1000 ml min^{-1} are show in Figure 11. The 1000 ml min^{-1} flow rate was chosen because in the column system used, it mimics the superficial gas feed velocities desired in actual field applications. In the column experiments, at a relatively high gasoline concentration, (*e.g.* 65 g gasoline (m^3 of inlet gas stream)$^{-1}$) the volumetric productivity is approximately 1200 g gasoline (m^3 bed medium)$^{-1}$ hr^{-1}.

Data from the field scale prototype biofilter are similar to the results obtained in the laboratory. Figure 12 shows hydrocarbon removal in terms of volumetric productivity. These data represent the lower end of the inlet gasoline concentrations tested in the laboratory but volumetric productivity values are comparable to those obtained with the continuous flow laboratory columns at similar inlet concentrations.

Figure 13 depicts total BTEX removal as a function of BTEX inlet flow rate. These data show that under the operating conditions employed, approximately 50-

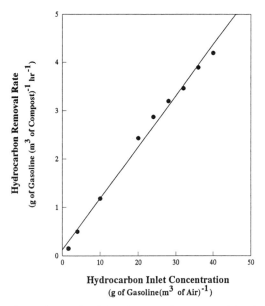

Figure 10. Volumetric productivity for gasoline removal by bed medium in serum vials at 30°C.

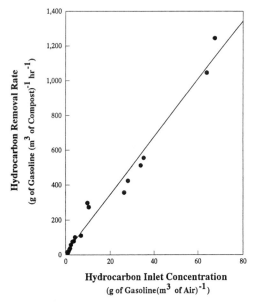

Figure 11. Volumetric productivity for gasoline removal by bed medium in continuous flow columns at 30°C.

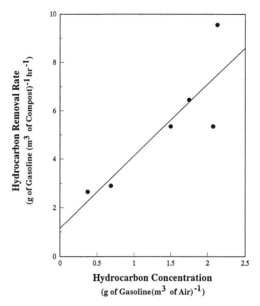

Figure 12. Volumetric productivity for total gasoline hydrocarbon removal by the field scale prototype biofilter.

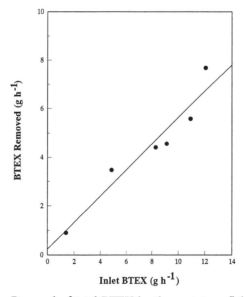

Figure 13. Removal of total BTEX by the prototype field biofilter.

55% of the BTEX compounds were removed. It should be noted however, that removal of toluene, ethylbenzene and xylene were typically > 80% while benzene, as predicted by the laboratory data, exhibited much lower removal rates of 10-15%.

The reaction rates of the biofilter systems tested at lower gasoline concentrations are probably considerably below those theoretically achievable due to substrate (*i.e.* gasoline vapor) limitations versus limitation related to insufficient microbial activity. The volumetric productivity data strongly indicate that to increase the reaction rates at working (*i.e.* lower) gasoline concentrations, optimization of biofilter design and/or the composition of the bed medium to increase contact between the gas stream and the microbes may be necessary. If the biofilter design/bed medium composition can be optimized while maintaining current or even higher levels of biological activity per unit bed volume, the gasoline removal rates could be substantially increased. Clearly, however, the first step to achieving increased gasoline removal rates is to develop a biofilter system in which increased gasoline transport to the microbes takes place.

Acknowledgments

This work was supported by the U. S. Department of Energy, Office of Technology Development, under DOE Idaho Field Office Contract DE-AC07-76IDO1570 and by EG&G Rotron.

Literature Cited

1. Apel, W. A.; Dugan, P. R.; and Wiebe, M. R. Use of methanotrophic bacteria in gas phase bioreactors to abate methane in coal mine atmospheres. *FUEL.* **1991,** 70, pp. 1001-1003.
2. Apel, W. A.; Dugan, P. R.; and Wiebe, M. R. Influence of kaolin on methane oxidation by *Methylomonas methanica* in gas phase bioreactors. *FUEL.* **1992,** 71, pp. 805-808.
3. Apel, W. A.; Dugan, P. R.; Wiebe, M. R.; Johnson, E. G.; Wolfram, J. H.; and Rogers, R. D. Bioprocessing of the Environmentally Significant Gases and Vapors Using Gas Phase Bioreactors: Methane, Trichloroethylene, and Xylene; Tedder, D. W.; and Poland, F. G., Eds.; ACS Symposium Series 518: Emerging Technologies in Hazardous Waste Management III; American Chemical Society: Washington, DC, 1993; pp. 411-428.
4. Bohn, H. L. Soil and compost filters of malodorous gases. *J. Air Poll Control Assoc.* **1975,** 25, p. 953.
5. Kampbell, D. H.; Wilson, J. T.; Read, H. W.; and Stockdale, T. T. Removal of Volatile Aliphatic Hydrocarbons in a Soil Bioreactor. *J. Air. Poll. Control Fed.* **1987,** 37, pp. 1238-1240.

6. Douglass, R. H.; Armstrong, J. M.; and Korreck, W. M. Design of a Packed Column Bioreactor for On-Site Treatment of Air Stripper Off Gas. In *On-Site Bioreclamation Processes for Xenobiotic and Hydrocarbon Treatment.* Henchee, R. E. and Olfenbuttel, R. F., Eds.; pp. 209-225.
7. Miller, D. E.; and Canter, L. W. Control of Aromatic Waste Air Streams by Soil Bioreactors. *Environ. Prog.* **1991,** 10, pp. 300-306.
8. Leson, G.; and Winer, A. M. Biofiltration: an innovative air pollution control technology for VOC emissions. *J. Air Waste Manage. Assoc.* **1991,** 41, pp. 1045-1054.
9. Travis, C. C.; and Macinnis, J. M. Vapor Extraction of Organics from Subsurface Soils: Is it Effective? *Environ. Sci. Technol.* **1992,** 26, pp. 1885-1887.
10. Johnson, P. C.; Stanley, C. C.; Kemblowski, M. W.; Byers, D. L.; and Colthart, J. D. A Practical Approach to the Design, Operation, and Monitoring of In Situ Soil-Venting Systems. *GWMR,* **1990,** Spring, pp. 159-178.
11. U. S. Environmental Protection Agency. In Situ Soil Vapor Extraction Treatment. EPA/540/2-91/006. U. S. EPA Cincinnati, OH. 1991.
12. Boethling, R. S.; and Alexander, M. Effect of concentration of organic chemicals on their biodegradation by natural microbial communities. *Appl. Environ. Microbiol.* **1979,** 37, pp. 1211-1216.
13. Leahy, J. G.; and Colwell, R. R. Microbial degradation of hydrocarbons in the environment. *Microbiol Rev.* **1990,** 54, pp. 305-315.
14. Riser-Roberts, E. Bioremediation of Petroleum Contaminated Sites. CRC Press, Boca Raton, FL. 1992; Vol. 1, pp. 59-64.
15. Perry, R. H.; and Chilton, C. H. Chemical Engineers' Handbook. 5th ed., McGraw-Hill Book Co., New York, NY, 1973; pp 16-12 - 16-18.
16. English, C. W.; and Loehr, R. C. Removal of organic vapors in unsaturated soil. Proceedings of Conference on Petroleum Hydrocarbons and Organic Chemicals in Groundwater: Prevention, Detection and Restoration. Special Issue of Groundwater Management. National Well Water Association. Pages 297-308, 1990.

RECEIVED December 1, 1993

Chapter 10

Production of Low-SO$_2$-Emitting, Carbon-Based Fuels by Coagglomeration of Carbonaceous Materials with Sulfur Sorbents

Abdul Majid, C. E. Capes, and Bryan D. Sparks

Institute for Environmental Chemistry, National Research Council of Canada, Ottawa, Ontario K1A 0R9, Canada

A liquid phase agglomeration technique is being developed to incorporate sulphur sorbents into carbonaceous fuels. This paper reports a summary of the work describing a number of case studies, based mainly on the work from the authors' laboratory, to illustrate the diverse applications of the technique. Athabasca petroleum cokes obtained from Suncor delayed coking and Syncrude fluid coking operations and a high sulphur Nova Scotia coal were used for these tests. Static combustion tests at 850°C were carried out in a muffle furnace and compared with results found for a bench scale fluidized bed unit at the same temperature.

It has been demonstrated that greater utilization of the sulphur sorbent can be achieved by cogglomerating the fuel with sulphur sorbents, providing an environment in which there is intimate contact between fuel and sorbent compared with systems in which the sorbent is added separately to the combustion bed. Coagglomeration of petroleum cokes and lime (Ca:S molar ratio of 1:1), resulted in a percent sulphur capture of at least, 30% to over 80% depending upon the source of carbon and the test conditions (fluid bed vs muffle furnace).

Oily sludges and organic wastes are produced by a number of industries, particularly those related to the recovery and processing of petroleum. These wastes pose several challenges. Traditional sludge disposal methods, involving concentration by impoundment followed by land filling or land farming, are meeting with increasingly stringent regulations. Further treatment of the wastes and reduction of volume and recycle are being encouraged and legislated. Such treatment may range from separation of constituents into higher value products, such as the separation of oil or other organic components from mineral (ash forming) impurities and water, to

0097–6156/94/0554–0160$08.00/0
Published 1994 American Chemical Society

stabilization of impurities to prevent leaching or to reduce emissions during combustion.

Liquid phase agglomeration (LPA) has the potential to play a major role in waste treatment processes (*1-2*). These range from the problem of waste tailings in the minerals and resource industries, to the disposal of organic sludges from petroleum production and refining operations to removal of sulphur impurities from coal or coke in order to mitigate the formation of acid rain. This process, under development at the National Research Council of Canada's Institute for Environmental Chemistry, relies on selective wettability, between mixtures of liquids and solids of different hydrophobic/hydrophillic character, to effect a separation of the components. Normally, a hydrophobic solid, such as ground coal or coke, is agitated vigorously with oil or oily sludge. During this process the oil is selectively adsorbed by the coal or coke to form a liquid film on the particle surfaces. Continued agitation of the mixture brings the oiled adsorbent particles into repeated contact with each other, resulting in flocculation of solids through the formation of interparticle liquid pendular bonds. These agglomerates, comprised of oil and hydrophobic carbon may then be separated from the aqueous phase using screens, cyclones or by degree of avidity for air bubble attachment, as in flotation. The type of separation scheme selected is governed by the degree of agglomeration achieved; this may range from weak floccules to densified, spherical agglomerates in which the voids are essentially saturated with oil. A more detailed description of the broad application of this process has been published in a number of review articles (*3-5*). Where water and hydrophillic solids are the contaminating phases in an oil based waste, then a hydrophillic solid adsorbent such as fine sand can be used as a collector for these components, leaving a clean oil (*6*). Similarly, emulsions can also be treated depending on whether they are water-in-oil or oil-in-water. It is usually more desirable to select an adsorbent which will preferentially collect the minor component in the material to be treated. A number of oily wastes have been effectively cleaned using solid carbonaceous materials for the recovery of residual oil (*7-10*). As well as collecting insoluble, oily contaminants a carbonaceous agglomerating solid may also adsorb most (80-90%) of the soluble organic species in a water based sludge. Consequently it is possible that such sludges can be cleaned effectively enough to allow direct sewerage of the treated water.

The cleaning of oily sludges by liquid phase agglomeration, using hydrophobic solid collectors such as ground coal or coke, has an added advantage in that not only is the oil adsorbed by the collector, but beneficiation of the collector, with respect to non-carbonaceous matter and pyrite also occurs during the process. However, organic sulphur will remain in the coal or coke matrix. This makes it imperative to include some form of desulphurization in any combustion system using this material as fuel.

The combustion of these agglomerates, with limestone addition, such as in a fluidized-bed reactor, could be one way to achieve the required reduction in sulphur dioxide emissions. However, it has been demonstrated (*11*) that this approach requires relatively high calcium to sulphur mole ratios, even with ash recycle, to produce acceptable reductions in sulphur dioxide emissions. Also, this technique precludes the use of finer sorbent particles that are known to be more efficiently utilized for sulphur capture. A major problem with such fine particles is that they are easily entrained in exhaust gases and thereby removed from the system. Incorporation of the sorbent into a pellet overcomes the problem of elutriation loss and allows finely ground particles to be used. The present research was designed to determine the feasibility of introducing finely divided sulphur dioxide capture agents into high sulphur fuels so that they will be more uniformly distributed and intimately contacted with other components within the fuel-sorbent agglomerates. On combustion the sorbent is expected to be more effective compared to a system in which it is added separately. Athabasca petroleum cokes, obtained from Suncor delayed coking and Syncrude fluid coking operations and a high sulphur Nova Scotia coal were tested for this purpose. This paper summarizes the work carried out with these materials.

1. Oily Waste Treatment

Recovery of Residual oil. The Hot Water Process for the separation of bitumen, from surface mined Athabasca oil sands, generates large amounts of oil-contaminated sludge. This sludge does not consolidate beyond about 35 w/w% solids and must be stored behind man-made dykes. Substantial quantities of bitumen and naphtha are lost to the ponds. For example, at the Suncor plant, pond #1 has a bitumen and naphtha content such that 2000 m^3/day could be extracted over a period of ten years. Consequently there is a strong economic as well as an environmental incentive to treat the sludge in these ponds.

Coking processes used in the upgrading of Athabasca oil sands bitumen, to form a synthetic crude oil, produce approximately 4000 tpd of coke. This coke is rather intractable as a fuel, being high in sulphur, low in volatiles, difficult to grind and having some relatively unreactive carbon forms, (12). Because of serious environmental and potential corrosion problems associated with the combustion of this coke, its use as a boiler fuel has been limited and a significant portion of the material is being stockpiled as a waste product. However, oil sands coke with a calorific value of about 33 MJ kg^{-1} would be an attractive boiler fuel if it could be desulphurized economically. This waste material is also an ideal adsorbent for the selective collection of oil from the ponds. A series of small scale bench tests (9) have been carried out using a laboratory high speed blender. Four main process variables were initially selected: blending time (min), amount of coke (wt. % of sludge), coke particle size (100% passing a given size) and the amount of dilution water (mls.), added to allow the suspension to be mixed properly. Blending was continued until the selected mixing time was completed. The treated slurry was then dumped onto a 100 mesh screen and the larger coke agglomerates separated from the remainder of the solids. After washing the agglomerates were analyzed for all components.

Results from these tests demonstrated that the amount and particle size of the oil adsorbent were the most important factors in achieving high oil recovery. When recovery of the organics was plotted against total available hydrophobic surface (total weight of coke times specific area) it was found that all the data fell close to a common curve as shown in Figure 1. An initial rapid rise in recovery upon addition of coke gradually tails off to a plateau, at about 85% oil collection, which represents the maximum attainable recovery under the given set of mixing conditions. In some cases, insufficient oil may be present to allow the formation of granules large enough to be separated by screening. In this situation bubble flotation is probably the best alternative for separation of the oil/coke flocs/agglomerates.

Removal of the oil component from the sludge usually results in more rapid and complete settling and compaction of the previously stable sludge. When this occurs it is possible to recycle a greater proportion of the pond water to the Hot Water Process. Consequently the tailings pond could be reduced in size.

Coagglomeration of Sulphur Dioxide Sorbents. The agglomerates obtained from the treatment of oily wastes by liquid phase agglomeration have potential use as an ancillary fuel. However, the heavy oils bitumens, petroleum cokes and coals, either present in the waste or added for treatment, are often high in sulphur. Thus, on combustion the emissions of sulphur dioxide may be above acceptable levels. The development of combined fuel-sorbent pellets or briquettes for use as a sulphur dioxide control method has been reported to give superior sulphur dioxide emission control during combustion, (13-14). It has already been demonstrated that the Athabasca oil sands bitumen; which is present in oil sands waste streams as a residual oil, is a good wetting agent for both the hydrophillic sulphur dioxide sorbent and hydrophobic coke. Consequently it is possible to incorporate sulphur dioxide adsorbents, such as finely divided limestone, into the agglomerates obtained during the treatment of oily sludge, (15-20). Syncrude refinery coke was used to collect over 90% of the residual bitumen from a sample of oil sands fine tailings (Suncor), according to the liquid phase agglomeration procedure reported previously (8). The discrete coke-oil agglomerates, of approximately 1 mm size, obtained from these tests, were then successfully coagglomerated with limestone corresponding to a Ca: S molar ratio of 1.2:1. This operation was made possible by the powerful collecting properties of bitumen and allowed the advantageous use of smaller and more active sulphur sorbent particles in fluid bed combustion, by binding them tightly within larger coal agglomerates. This approach reduces the possibility of elutriation of the SO₂ sorbent particles from the bed and, higher sorbent utilization efficiencies can be obtained for the coagglomerated fuel, compared to those systems in which coarser sorbent particles are added separately to the fluid bed. Combustion tests at 850°C in a bench scale fluidized bed reactor indicated over 90% sulphur capture (unpublished data) for these samples.

Similarly, it is possible to use a wet agglomeration process to produce a conglomerate of coke and sorbent; samples from both Suncor and Syncrude operations were successfully coagglomerated with either limestone, lime or hydrated lime using bitumen as the binding liquid. During combustion of the coke-bitumen-sorbent coagglomerates, sulphur dioxide capture was found to depend mainly on the calcium to sulphur mole ratio, the combustion temperature, and the type of coke. Moisture content of the agglomerates did not have any significant effect on sulphur capture. The combustion tests on coke-sorbent agglomerates were carried out in either a bench scale fluidized bed apparatus at 850°C or a muffle furnace at 460-1000°C. Figure 2 shows the sulphur dioxide emissions as a function of Ca:S ratio

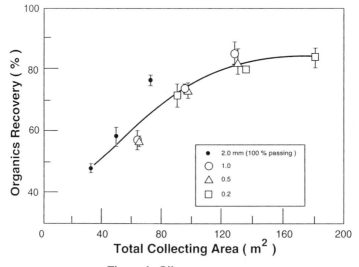

Figure 1. Oil recovery curve

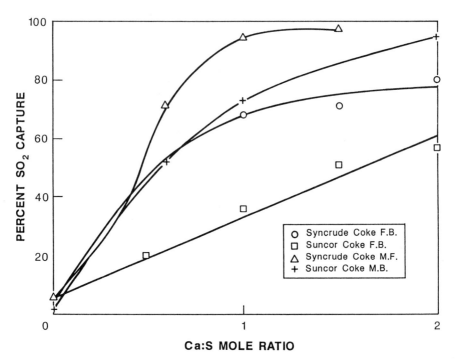

Figure 2. Effect of Ca:S molar ratio on the retention of SO_2 by lime

in the coagglomerated material, (*18*). Test results indicated sulphur capture of over 60 wt % for Syncrude coke and over 30 wt % for Suncor coke at a calcium to sulphur molar ratio of 1:1. Figure 3 describes the comparative SO_2 capture efficiencies of the three sorbents. With the observed scatter of results, no significant difference in sulphur capture between the three sorbents could be detected. However, the fact that the limestone used in this series of tests gave results as good as lime or hydrated lime has considerable economic significance. The cost ratio of limestone on a molar basis may vary from 2 to 4, depending on the transportation distance. Even the costs for transportation and handling of limestone tend to be lower than for lime because limestone can be transported in open trucks. Also, limestone is readily available in the Athabasca region of Alberta.

According to the findings of Schneider and George (*21*), the presence of calcium also has a beneficial effect on the acid leaching of nickel and vanadium from coke ash. Hence, coagglomeration of coke with calcium compounds will have the added advantage that the ash from the burnt agglomerates will be more suitable for heavy-metal recovery as a byproduct of the combustion process.

2. Fine Coal Recovery

Agglomeration techniques have been used to recover coal from washery wastes and tailings ponds (*1*). Compared to flotation methods the process has been shown to be particularly effective for very fine materials. During recovery the coal is also beneficiated with respect to ash forming components and inorganic sulphur compounds, such as pyrites.

Separation of Pyrite. Two main problems are associated with the removal of pyritic sulphur from coals. The first is that pyrite is often in a fine state of dissemination, which requires grinding to a very fine size to accomplish liberation (*22*). The second problem is caused by the similar surface chemical characteristics of coal and pyrite, which complicates the use of separation methods based on surface wettability (*23*).

The problem of fine particle size for pyrite liberation can be partly overcome by the selective agglomeration method. With very fine grinding and the aid of a number of possible pyrite depressants under acidic, neutral and alkaline conditions it was possible (*24*) to remove 50% of the pyrite present in a high volatile bituminous coal. This agrees with the general difficulty experienced in removing pyrite by surface wetting differentiation as is used in both oil agglomeration and froth flotation.

Recent work on coal beneficiation by liquid phase agglomeration has shown that aging of agglomerated coal could be advantageous in the beneficiation of pyritic sulphur (*15*). A sample of bituminous thermal coal from the Prince Mine, Cape Breton Island, Nova Scotia, was agglomerated with No. 4 fuel oil and stored in the laboratory for two months. This treated coal was then redispersed in water and reagglomerated with Athabasca oil sands bitumen; overall considerable further beneficiation with respect to non-carbonaceous matter and pyritic sulphur was achieved. In the second agglomeration step a reduction in pyritic sulphur of 80% was achieved compared to 40% in the original treatment. Ash content of the final agglomerates was 4±0.1% compared to 10.9% for the initial agglomerates and 19 % for the feed material. It is suspected that aging of the original coal agglomerates had resulted in selective oxidation of pyrite surfaces, thereby facilitating removal of this component because of the more hydrophillic nature of the oxidized particles. This is consistent with the successful removal of pyrite from coal using a combination of bacterial treatment followed by oil phase agglomeration, (25-26). Bacteria are known to oxidize pyrite, rendering the particle surfaces more hydrophillic; during this process surfactants are produced that could also affect the wettability of the components. However, coal surfaces, protected by an oil coating, are not significantly affected by either oxidation or surfactants.

Coagglomeration with Sulphur Dioxide Sorbents. Physical coal cleaning methods can remove only inorganic forms of sulphur, leaving the organic form in the coal matrix. Therefore, in addition to precombustion cleaning, inclusion of some form of desulphurization in any combustion system using this material as a fuel may also be necessary. Our earlier work on petroleum coke had demonstrated that coagglomeration of coke and sorbent as well as beneficiation, with respect to ash content of the coke can be achieved in a single step using a wet agglomeration process. Therefore, further work was carried out to determine the possibility of coagglomerating sulphur sorbents with coal as a means of reducing sulphur dioxide emissions during combustion. In a series of tests a run-of-mine sample of Nova Scotia coal was coagglomerated with limestone during the primary cleaning stage, using bitumen as binder. Static combustion tests at 850°C were carried out in a muffle furnace and compared with results found for a bench scale fluidized bed unit at the same temperature. The results are shown in Figure 4. In both cases, sulphur capture of over 60% was obtained at a calcium to sulphur molar ratio of 1:1.
Attempts to use LPA techniques to prepare a composite from the aged, agglomerated coal, limestone and bitumen were unsuccessful. It appears that the surfactants produced by pyrite consuming bacteria may prevent bonding of hydrophillic limestone with bitumen. Instead, the agglomerated, weathered coal was pelleted with varying proportions of limestone using a die and press. Combustion tests on these pellets were carried out in a muffle furnace at 850°C. The results, shown in Figure 5, also contain data for the run-of-mine coal/limestone agglomerates. This plot indicates similar sulphur retention results for the two types of agglomerates, suggesting that they both withstood the combustion conditions equally well.

Comparative Sulphur Capture Efficiencies

The data for cokes and coal has been combined in Figure 6 for comparison purposes. This is a plot showing the levels of SO_2 emissions obtained for both blank and lime containing agglomerates of Prince coal, Suncor and Syncrude cokes. USA and Canadian SO_2 emission standards are also shown on this plot. This Figure clearly illustrates that a Sulphur capture capacity of more than 80% may be needed to burn these fuels to meet SO_2 emission standards. The data for coal/coke lime agglomerates plotted in the Figure shows that coagglomeration of fine sulphur

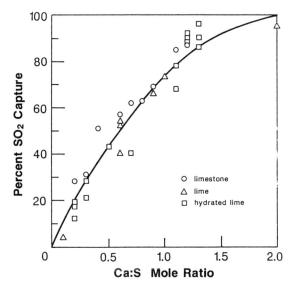

Figure 3. Comparative SO_2 capture efficiencies of various sorbents for Suncor coke

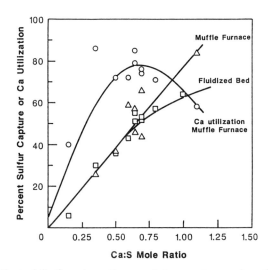

Figure 4. Effect of Ca:S molar ratio on sulphur capture and on Ca utilization

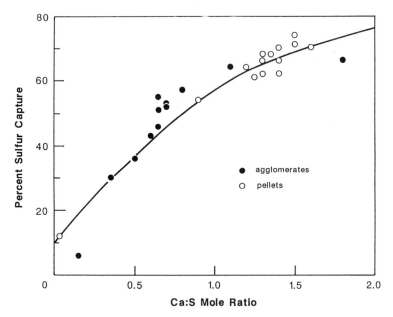

Figure 5. Effect of Ca:S molar ratio on sulphur capture for coal/limestone
agglomerates/compacts during combustion in a muffle furnace

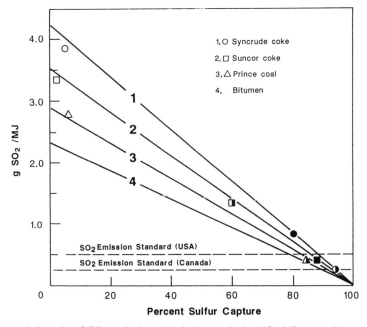

Figure 6. Levels of SO_2 emission: blank (open circles); fuel-lime agglomerates
(half closed symbols, static bed results, closed symbols, FBC). Ca:S
molar ratios: coke, 2; coal, 1.1

sorbents in amounts representing a Ca:S molar ratio in the range of 1 to 2 could result in the reduced SO_2 emissions necessary to meet proposed standards.

Conclusion

A liquid phase agglomeration process has been developed to incorporate finely divided sulphur dioxide capture agents into a variety of carbonaceous fuels. The resulting agglomerates had improved combustion characteristics and lower sulphur dioxide emissions compared with those for the original fuels. The decrease in sulphur dioxide emitted on combustion was found to depend on the calcium to sulphur mole ratio, source of fuel, and conditions of testing. The coal/coke-bitumen-limestone coagglomeration results suggest that coagglomeration of fine sulphur dioxide sorbents in amounts representing a Ca:S molar ratio in the range of 1 to 2 could result in the reduced SO_2 emissions necessary to meet USA as well as Canadian proposed emission standards. This chapter was issued as NRCC No. 35767.

Literature Cited

1. Capes, C. E. In, "Challenges in Mineral Processing"; Editors, K. V. S. Sastry and M. C. Fuerstenau, Society of Mining Engineers Inc., Littleton, Colorado, **1989**.
2. Sparks, B. D., Capes, C. E., Hazlett, J. D. and Majid, A., Proc. Int. Symposium on tailings and effluent Management, Halifax, August 20-24, **1989**. Pergamon Press, Toronto.
3. Sirianni, A. F., Capes, C. E., and Puddington, I. E., Can. J. Chem. Eng., **1969**, Vol. 47, pp. 166-170.
4. Capes, C.E. Can. J. Chem. Eng. **1976**, Vol. 54, pp. 3-12.
5. Sparks, B. D., Farnand, J. R., and Capes, C. E., J. Sep. Process Tech., **1982**, Vol. 3, No. 1, pp. 1-15.
6. Capes, C. E., Coleman, R. D., Germain, R. J. and McIlhinncy, A. E., Proceedings Coal and Coke Sessions, 28th Can. Chem. Eng. Conf., Chemical Institute of Canada, Ottawa, **1978**, pp. 118-131.
7. Majid, A. and Ripmeester, J. A., J. Sep. Process Technol. **1983**, Vol. 4, pp. 20-22.
8. Majid, A.; Ripmeester, J.A.; Sparks, B.D., Proceedings of the 4th International Symposium on Agglomeration, **1985**, pp 927-935.
9. Kumar, A., Sparks, B. D. and Majid, A., Sep. Sci. Technol., **1986**, Vol. 21, pp. 315-326.
10. Majid, A. and Sparks, B. D., Proceedings of Tar Sand Symposium, Jackson, Wyoming, **1986**, paper 7-2.
11. Anthony, E. J., Desai, D. L. and Friedrich, F. D., "Fluidized bed combustion of Suncor coke", CANMET Report No. EPR/ERL 81-27; **1981**, CANMET, Ottawa, Canada.
12. Friedrich, F. D., Lee, G. K., Desai, D. L. and Kuivalainen, R., ASME Paper No. 82-JPGC-Fu-3, **1982**
13. Conkle, H. N., Dawson, W. J. and Rising, B. W., Proc-Inst. Briquette. Agglom., Biennial Conf. **1983**, Vol. 18, pp. 33-54.

14. Sidney, M. H., "Method for removal of sulphur from coal in stoker furnaces", U. S. A. Patent No. 4,173,454, **1979**.
15. Majid, A., Sparks, B. D., Capes, C. E. and Hamer, C. A., Fuel, **1990**, Vol. 69, pp.570-574.
16. Majid, A., Clancy, V. P. and Sparks, B. D., ACS, Div. of Fuel Chem. Preprints, **1987**, Vol., 32, #4, pp.412-432.
17. Majid, A., Clancy, V. P. and Sparks, B. D., Energy & Fuels, **1988**, Vol. 2, pp. 651-653.
18. Majid, A., Sparks, B. D. and Hamer, C. A., Fuel, **1989**, Vol., 68, pp. 581-585
19. Majid, A., Sparks, B. D. and Hamer, C. A., Proc. 4th UNITAR Int. Conf. Heavy Crude and Tar Sands, **1989**, Vol., 5, pp. 397-404.
20. Majid, A., Sparks, B. D., Capes, C. E. and Hamer, C. A., Proc. 21st Biennial Conf. of the Institute for Briquetting and Agglomeration, **1989**, Vol. 21, pp. 403-414.
21. Schneider, L. G. and George, Z. M., Ext. Metall. 81, pap. Symp., **1981**, pp. 413-420.
22. Greer, R. T. In, " coal desulphurization: Chemical and physical methods", Editor, T. D. Wheelock, ACS Symposium series, 64, American Chemical Society, Washington, D. C., **1977**, pp. 3-15.
23. Coal preparation, Leonard, J. W. and Mitchell, Eds.; 3rd ed., AIME, New York, **1968**.
24. Capes, C. E., McIlhinney, A. E., Sirianni, A. F. and Puddington, I. E., In, "upgrading pyritic coals", Can. Inst. Mining Met. Bull., 1973, Vol. 66, pp. 88-91.
25. Kempton, A. G., Nayera Moneib, R. G. L. McCready and C. E. Capes, Hydrometallurgy, **1980**, Vol. 5, pp. 117-125.
26. Schlick, S., Narayana, M. and Kevan, L., Fuel, **1986**, Vol. 65, pp. 873-876.

RECEIVED December 1, 1993

Chapter 11

Removal of Nonionic Organic Pollutants from Water by Sorption to Organo-oxides

Jae-Woo Park and Peter R. Jaffé

Water Resources Program, Department of Civil Engineering and Operations Research, Princeton University, Princeton, NJ 08544

Anionic surfactant monomers can be adsorbed onto mineral oxides from the aqueous phase if the pH of the solution is below the oxide's zero point of charge (ZPC). The oxide with anionic surfactant sorbed to it is called organo-oxide, and it acts as a sorbent for nonionic organic pollutants, since these pollutants will partition between water and the organic phase of the sorbent. The advantage of this sorbent is that, unlike activated-carbon, it can be regenerated in-situ. Batch and column experiments were done to demonstrate the use of an organo-oxide for the treatment of water contaminated with a nonionic organic pollutant. The results from column experiments matched well with theoretical predictions based on parameters obtained from batch experiments.

Oxides exhibit surface charges in the aqueous environment. These surface charges are quite pH-dependent because hydrogen (H^+) and hydroxyl (OH^-) ions are potential determining ions on mineral oxides. The ZPC of an oxide is the pH at which the solid surface charges from all sources are zero. Oxides take on positive charges when the pH is lower than their ZPCs, while they take on negative charges when the pH is higher than their ZPCs (1).

Surfactants are commonly classified as anionic, cationic, amphoteric, and nonionic depending on the charges of the hydrophilic parts of their monomers. Anionic surfactants can be adsorbed strongly onto oxides in an acid environment (pH below ZPC), while they can be desorbed from oxides in a basic environment (pH above ZPC). These sorbed surfactants provide an organic phase into which nonionic organic substances can be partitioned (2-5).

Several researchers have investigated the combination of oxides and anionic surfactant as a synthetic sorbent. Recently, Valsaraj (2) has described the partitioning of three volatile organic compounds between the aqueous phase and two surfactant aggregates adsorbed onto aluminum oxide in closed systems. Holsen et al. (5) have also shown that sparingly soluble organic chemicals can be removed from the aqueous phase using anionic surfactant-coated ferrihydrite in batch experiments. Park and Jaffé (6) have illustrated the technical feasibility of using organo-oxides in a continuous-flow water-treatment process for the removal of nonionic organic pollutants. Unlike for the activated carbon process, the column containing the anionic surfactant-treated aluminum oxide can be

0097–6156/94/0554–0171$08.00/0

regenerated in-situ by first increasing the pH of the solution in order to desorb the surfactant and the nonionic organic pollutant, followed by the readsorption of surfactants at a lower pH. As learned from our previous work (6), during the regeneration process a relatively large volume of flushing solution was required to desorb the surfactant and nonionic organic pollutant from the oxide.

The objective of this study was to: (1) further investigate the interaction between organo-oxides and nonionic organic pollutants in batch experiments, (2) modify the continuous-flow column experiments and their regeneration process in order to increase the regeneration efficiency, and (3) develop a model that can be used to predict the dynamics in an oxide/surfactant/nonionic organic pollutant system, and that can be applied to predict the sorption capacity of the organo-oxide column, as well as the amount of flushing solution required to achieve a desired level of regeneration.

Theoretical Considerations

Surfactants are known to exist in monomeric state if the concentration is below the critical micelle concentration (CMC). Above the CMC, if an additional mass of surfactant is added to a solution, the concentration of monomers remains constant while the excess surfactant leads to the formation of micelles (7). In addition to these two aqueous phases of the surfactant, for the system described here, there is a third phase, which is the surfactant sorbed onto the oxide particles. A general expression for the mass balance of surfactant can be written as

$$T^s = S^s + M_1^s + M_2^s \qquad (1)$$

where T^s is the total mass of surfactant added, S^s is the mass of surfactant adsorbed onto the oxide particles which is also referred to as hemimicelles, M_1^s is the mass of surfactant dissolved in water in micellar form, and M_2^s is the mass of surfactant dissolved in water in monomeric form.

Kile and Chiou (8) presented a general expression for the solubility enhancement of a solute by surfactants in terms of the concentrations for monomers and micelles and the corresponding solute partition coefficients. Since there is an additional surfactant phase in our system, a general expression for the mass balance of a solute in the presence of the three surfactant phases would be

$$T^c = S^c + M_1^c + M_2^c + C^c \qquad (2)$$

where T^c is the total mass of the solute added, S^c is the mass of the solute sorbed into the organic phase of the organo-oxide, M_1^c is the mass of the solute sorbed into the micelles in aqueous phase, M_2^c is the mass of the solute associated with the monomers in the aqueous phase, and C^c is the mass of the solute dissolved in water.

It is shown in the literature (8-11) that the solubility of nonionic organic compounds in water is essentially not affected by the surfactants when the surfactant's concentration is below CMC, except for some extremely water-insoluble organic solutes (e.g., DDT). Since in this work we will not consider extremely water-insoluble compounds, we will assume that M_2^c is negligible, then the original mass balance eqn. (2) is reduced to

$$T^c = S^c + M_1^c + C^c \qquad (3)$$

Two partition coefficients can be defined that describe the partitioning process of the solute between the different organic phases and the aqueous phase as

$$K_p = C_p / C_e \qquad (4)$$
$$K_m = C_m / C_e \qquad (5)$$

where K_p is the partition coefficient of the solute between the organo-oxide and the aqueous phase (L/Kg), C_p is the concentration of the solute sorbed onto the organo-oxide (mg/kg of organo-oxide), and C_e is the concentration of the solute dissolved in water. K_m is the partition coefficient of the solute between the micellar phase and the aqueous phase (L/Kg) and C_m is the concentration of the solute in micellar phase (mg/Kg of micelles). Using equations (4) and (5) to express eqn. (3) in terms of aqueous concentrations, we obtain

$T^c = C_p$ * (mass of organo-oxide) + C_m * (mass of micelles) + C_e * (volume of water)

$$T^c = K_p * C_e * \text{(mass of organo-oxide)} + K_m * C_e * \text{(mass of micelles)} + C_e * \text{(volume of water).} \qquad (6)$$

As shown in eqn. (6), if K_p and K_m are known, the distribution of the solute between each phases in equilibrium can be predicted.

Experimental Section

The oxide used in this study was aluminum oxide (Al_2O_3), purchased from Aldrich Chemical Co. Particle sizes of the aluminum oxide are less than 150 mesh and the surface area of the oxide is 155 m^2/g. A total of 250 μCi of [^{14}C]carbon tetrachloride (specific activity equal to 4.3 mCi/mmol) was purchased from Du Pont NEN. [^{14}C]carbon tetrachloride (CCl_4) was mixed with nonradioactive carbon tetrachloride to yield net volume of 1.0 mL, resulting in a specific activity of 31.2 μCi/mmol. The anionic surfactant, Emcol CNP-60, was obtained from Witco Co, and its structure is shown in Table 1. Aluminum oxide and the anionic surfactant were used as received.

The adsorption of the anionic surfactant onto aluminum oxide in batch experiments was studied by placing 4 g of aluminum oxide and 56 mL of various concentrations of the surfactant solution in 60 mL glass centrifuge tubes with Teflon-lined caps. The organic carbon content of the organo-oxide was quantified in duplicate by Huffman Laboratories, Golden, CO. The sorption of carbon tetrachloride onto the anionic surfactant-treated aluminum oxide in batch experiments was studied by placing 1 g of the oxide, 14 mL of the surfactant solution, and appropriate volume of [^{14}C]carbon tetrachloride in 15 mL glass centrifuge tubes with Teflon-lined caps. In both cases, the samples were equilibrated in the dark at 20 °C, during which the tubes were rotated continuously for complete mixing. Even though kinetic experiments show that surfactant equilibrium between the aqueous and solid phases and solute equilibrium between the aqueous and organo-oxide phases were reached fast (4-5), a 48 hour incubation time was used to ensure that equilibrium was reached in all samples. After the incubation, the samples were centrifuged for 60 minutes at 650g (g = 9.81 m/s^2).

After the samples were centrifuged, the dissolved mass of either the surfactant or carbon tetrachloride was determined. Surfactant concentrations were determined by measuring the surface tension of the aqueous supernatant with a Fisher Scientific Model 21 surface-tensiometer that employs the du Nouy ring method. Surface tension values were taken when stable readings were obtained for a given sample, as indicated by at least two consecutive measurements having nearly the same value. A Packard Tri-Carb 1900CA liquid-scintillation analyzer was used to measure the radioactivity in the sample. The equilibrium solute concentration in the supernatant from the centrifuge tube was then determined by a standard curve relating disintegration per minute to aqueous concentration. Since the total mass of surfactant and solute added to the sample was known, the mass sorbed could be determined by subtracting the dissolved mass from the mass added. In order to determine if there was loss of the solute due to sorption of the solute onto the inside of the centrifuge tubes or due to volatilization, blank samples were prepared and handled in parallel with other samples in each set of the batch experiments of the sorption of carbon tetrachloride onto the organo-oxide. Blank samples consisted of water, surfactant, and radio-labeled carbon tetrachloride combined in a centrifuge tube without the oxide.

The continuous-flow water treatment system consisted of a glass column with an inner diameter of 1.5 cm and a length of 10 cm. The column was filled with 10 g of aluminum oxide. After packing the column with the oxide, several pore volumes of de-aired and de-ionized water were flushed through the column in both upward and downward directions to obtain fully saturated conditions. A schematic of the column setup is shown in Figure 1.

Table 1. Structure of Emcol CNP-60
Emcol CNP-60

$$\begin{array}{c} CH_3 \\ \diagdown \\ \diagup \\ CH_3 \end{array} CH - CH_2 - \underset{\underset{CH_3}{|}}{CH} - CH_2 - CH_2 - CH_2 - \underset{\bigcirc}{\langle \bigcirc \rangle} - O(CH_2CH_2O)_6CH_2 - CO_2H$$

Emcol NP-60

$$\begin{array}{c} CH_3 \\ \diagdown \\ \diagup \\ CH_3 \end{array} CH - CH_2 - \underset{\underset{CH_3}{|}}{CH} - CH_2 - CH_2 - CH_2 - \underset{\bigcirc}{\langle \bigcirc \rangle} - O(CH_2CH_2O)_6H$$

The product Emcol CNP-60 consists of approximately 70% Emcol CNP-60, ~18% Emcol NP-60, 10% H_2O, and ~1% inorganic salts.

To coat the aluminum oxide in the column with Emcol CNP-60 monomers, a solution of Emcol CNP-60 with a concentration of 14 g/L (CMC = 59 mg/L) was pumped into the column from top to bottom at a flowrate of 27 mL/hr. for 2 hours. After breakthrough of the surfactant was observed, and the organo-oxide was produced, water was pumped into the column from top to bottom at a rate of 27 mL/hr.. Before the water entered the column, carbon tetrachloride was added to the water stream with the aid of a syringe pump at a rate of 8.05 µL/hr. for 6 hours, resulting in a carbon tetrachloride concentration of 451 mg/L. The setup for carbon tetrachloride injection is shown inside the dotted box in Figure 1.

Finally, after breakthrough of carbon tetrachloride was observed, and in order to regenerate the column, a 2.5 N sodium hydroxide (NaOH) solution (pH 13.8) was pumped into the column to change the surface charges of the oxide in order to facilitate the removal of the anionic surfactant and the solute. The sodium hydroxide solution was pumped into the column in an upflow mode. After one pore volume (14 mL) was replaced with 2.5 N NaOH solution, the column was placed into a rotating tumbler for 24 hours. After this an additional pore volume was replaced with a fresh 2.5 N NaOH solution and the column was placed back into the tumbler for 10 minutes. This procedure was repeated twice. Each pore volume obtained in this manner was analyzed to determine the concentration of the surfactant and carbon tetrachloride using the methods described above.

Results and Discussion

a) Batch Experiments   Results from a screening test, in which several oxide/surfactant combinations were tested indicated that for the combinations tested, the most efficient one for the removal of carbon tetrachloride from water was aluminum oxide and Emcol CNP-60 (6). The organic carbon content of this organo-oxide is 9.1 % when the Emcol CNP-60 monomers were adsorbed from a non-buffered solution onto aluminum oxide surface up to the uptake capacity of the oxide.

Adsorption isotherms of the anionic surfactant onto aluminum oxide at several pHs are plotted in Figure 2. In low pH where there are more positive sites than negative sites on the oxide's surface, more surfactant monomers are adsorbed onto the oxide particles. As the mass of surfactant sorbed onto the oxide increases, resulting in an increased organic carbon content of the organo-oxide, the partitioning of the solute into the organo-oxide will also increase.

To quantify the two solute partition coefficients described above (K_p and K_m), a series of batch experiments was conducted without pH adjustment. Sorption isotherm of carbon tetrachloride onto the organo-oxide with a 7.1% organic carbon content is plotted in Figure 3, where the slope of the sorption isotherm is K_p for that specific organic carbon content. To calculate K_p for different values of the organic carbon content of the sorbent, a set of batch experiments was conducted and the result is shown in Figure 4, where the logarithms of K_p values and logarithms of organic carbon normalized K_p values are plotted as a function of the organic carbon content of the organo-oxide. Note that for the pH at which these experiments were conducted, essentially all of the surfactant is adsorbed onto the oxide, hence the surfactant concentration in solution is much smaller than CMC. As shown in Figure 4, organic carbon normalized partition coefficients between organo-oxide and dissolved phase is about the same as the octanol-water partition coefficient (K_{ow}) of carbon tetrachloride, which is 2.64. The magnitude of the partition coefficient and the linearity of the sorption isotherm indicate that the uptake of carbon tetrachloride by the organo-oxide used here is a partitioning process rather than an adsorption process. This is also in agreement with the results reported by Smith et al. (4) and Smith & Jaffé (12) who showed that organo-clays formed with long-carbon chain surfactants, similar in size to Emcol CNP-60, provide a partitioning media. A stronger uptake could be achieved through an adsorption process, for which one should select smaller surfactant molecules.

To obtain Km values, an adsorption isotherm of the surfactant onto the aluminum oxide and a series of sorption isotherms of carbon tetrachloride onto the organo-oxide were conducted at a pH of 13.8. (Figure 2 and 5). The pH was set to 13.8 because the regeneration of the column was performed at that pH. Five sets of sorption isotherms of the solute onto the sorbent were conducted for different quantities of excess surfactant added to the oxide. To each sample sufficient surfactant was added to form micelles in the aqueous phase after the oxide was fully coated with the surfactant. Two of these sorption isotherms of the solute onto the sorbent for the lowest and the highest of the five micellar concentrations are shown in Figure 5. The micellar concentration is defined here as the dissolved surfactant concentration in excess of CMC, or M_1^s/volume of the solution. The results show that as the micellar concentration of the surfactant becomes larger, the mass of the solute sorbed onto the sorbent (S^c) gets smaller. This is because more mass of the solute is sorbed into micelles (M_1^c) rather than onto the organo-oxide. Since only the total mass of the solute in the aqueous phase, which is the sum of mass of the solute sorbed into the micelles in the aqueous phase (M_1^c) and mass of

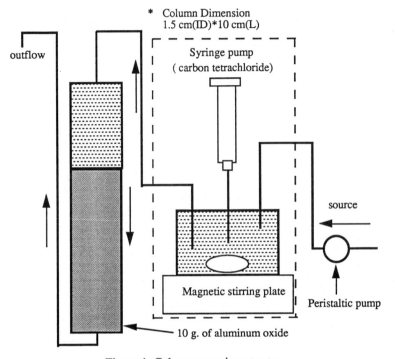

Figure 1. Column experiment setup.

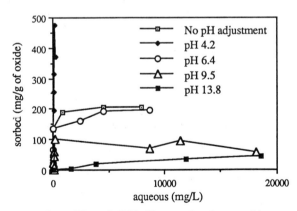

Figure 2. Adsorption of Emcol CNP-60 onto aluminum oxide at several pHs.

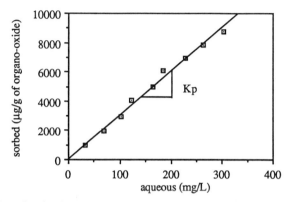

Figure 3. Sorption isotherm of carbon tetrachloride onto the organo-oxide with a 7.1% organic carbon content at non-buffered pH.

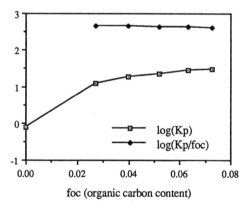

Figure 4. K_p(L/Kg) and organic carbon normalized K_p for carbon tetrachloride vs. organic carbon content of the sorbent.

the solute dissolved in water (C^c), could be determined with the analytical technique described above, the concentration of the solute dissolved in water (C_e) had to be estimated indirectly. The dissolved concentration (C_e) was estimated using eqn. (4), for which the concentration of the solute sorbed onto the organo-oxide (C_p) could be determined directly from these batch experiments. The corresponding K_p was estimated as follows: The mass of surfactant sorbed onto the oxide at a pH of 13.8, and therefore the oxide's organic carbon content, was determined from the results shown in Figure 2. Kp could be estimated for that specific organic carbon content from the results shown in Figure 4. Once the concentration of the solute dissolved in water (C_e) was obtained, the concentration of the solute in micellar phase (C_m) was determined by subtracting the mass of the solute dissolved in water (C_e) from the measured total mass of the solute in the aqueous phase. Knowing C_e and C_m, K_m could be estimated for each one of the five sorption isotherms of the solute onto the sorbent, using eqn. (5). The result is shown in Figure 6, where the logarithm of K_m is plotted as a function of the micellar concentrations. If Kp is normalized with respect to the mass of sorbed surfactant, one obtains Km for the hemimicelles on the oxide. The results presented in Figure 6 show that K_m values of micelles in the aqueous phase and calculated K_m values of hemimicelles on the oxide are quite similar, suggesting that the sorption characteristics of the monomers do not change much whether they exist as micelles or hemimicelles.

The results presented in Figures 4 and 6 were used to quantify K_p and K_m, based on which the results of the column experiments described in the next section were simulated.

b) Column Experiments A solution of Emcol CNP-60 was pumped into the laboratory-scale water treatment system described above. The results are shown in Figure 7, where C_i is the concentration of surfactant in the influent, C is the concentration of surfactant in the effluent, T_o is the hydraulic residence time, and T is the time since pumping was started. Run 1 and Run 2 are replicate experiments.

Subsequently, water containing carbon tetrachloride was pumped into the column until the concentration of the contaminant in the effluent was about 50 % of that in the influent. The results are shown in Figure 8, where C_i is the concentration of carbon tetrachloride in the influent and C is the concentration of the contaminant in the effluent.

Finally the regeneration of the column using a solution with a pH of 13.8 was done. As described above, the purpose was first to remove the anionic surfactant and the contaminant from the aluminum oxide, in order to regenerate the column with new surfactant monomers. The desorbed fractions of surfactant and carbon tetrachloride in each pore volume of flushing solution are shown in Figure 9 and 10. Run 1 and Run 2 represent the results from actual column experiments, while Run 1-theo. and Run 2-theo. represent the predicted results from the mass balance models described earlier and using the partition coefficients from two standard curves from the batch experiments (Figure 4 and 6). The values of Run 1-theo. and Run-2 theo. are calculated for each pore volume using the actual remaining mass of surfactant and carbon tetrachloride after the previous pore volume. As indicated in Figure 9 and 10, the results from the actual experiments and the predicted results from the mass balance models were matched well.

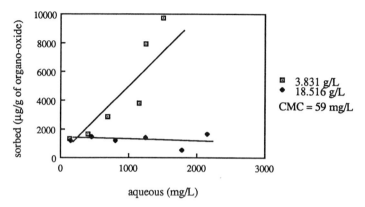

Figure 5. Sorption isotherm of carbon tetrachloride onto the organo-oxide at pH 13.8.

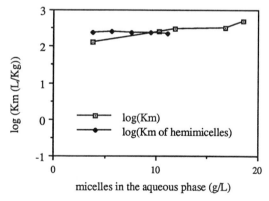

Figure 6. K_m(L/Kg) and K_m of hemimicelles in the sorbent for carbon tetrachloride vs. micelle concentration in the aqueous phase.

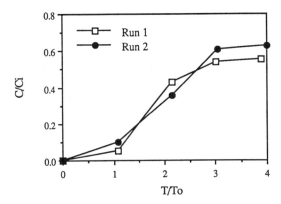

Figure 7. Breakthrough of the surfactant from the oxide column.

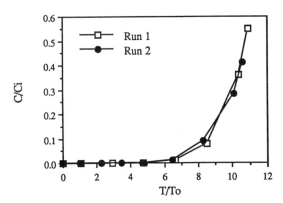

Figure 8. Breakthrough of carbon tetrachloride from the organo-oxide column.

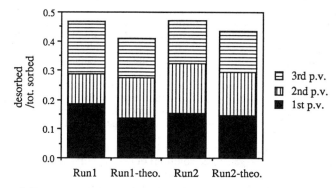

Figure 9. Recovery of the surfactant during the regeneration of the column.

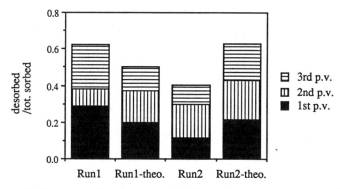

Figure 10. Recovery of carbon tetrachloride during the regeneration of the column.

Summary and Conclusions

Sorption of an anionic surfactant (Emcol CNP-60) onto aluminum oxide and the subsequent partitioning of (a nonionic organic substance) carbon tetrachloride was studied. Sorption of the anionic surfactant onto the oxide is strongly pH-dependent, and decreases as the pH increases. The partitioning of the nonionic organic pollutant onto this synthetic sorbent is linearly dependent on the concentration of the pollutant and is proportional to the adsorbed mass of the surfactant.

Treatment of water contaminated with carbon tetrachloride was investigated. For this purpose a column was first loaded with aluminum oxide , then the surfactant solution was pumped through the column to prepare the synthetic sorbent. The contaminated water was then pumped through the column until about 50 % of the concentration of carbon tetrachloride in the influent was observed in the effluent. In order to regenerate the column, a sodium hydroxide solution was pumped through the column to desorb the surfactant and the organic pollutant. The results from the actual regeneration process could be predicted well using mass balance models for the surfactant and for the solute, using partition coefficients determined from the batch experiments. Therefore, the methodology presented here can be used for the design and evaluation of such water treatment systems.

The concentration factor obtained in this work (volume of contaminated water treated/volume of regenerated solution) of carbon tetrachloride was 7 to 5. If more hydrophobic contaminants are present, a higher treatment efficiency will be accomplished, because the values for Kp will be significantly larger, which increases the volume of water that can be treated while the volume of water used in the regeneration is fairly constant. Furthermore, as discussed earlier, the selection of smaller surfactant molecules used in producing the organo-oxide may increase the sorption efficiency by orders of magnitude (4).

It is interesting to note that the mixing that was provided during the regeneration had a significant effect on increasing the desorption efficiency compared to that obtained previously (6). For this reason, the regeneration of a larger-scale column should be conducted in an upflow mode and the water should be pumped at a velocity that allows for an expansion of the bed and, therefore, better mixing.

Even though this organo-oxide column might not be as efficient as an activated carbon column in terms of bulk treatment of water, it has several advantages that might be of interest in specific applications: (1) it can be regenerated in-situ, (2) selective removal of a certain contaminant may be achieved if a specific surfactant that sorbs the contaminant selectively is used (13), and (3) the solute that is removed from water by the organo-oxide can be recovered if this is desired.

Acknowledgments

This research was funded by a grant from Carter-Wallace, Inc. The anionic surfactants and their chemical structures were provided by Witco Corporation.

References

(1)   Parks, G. A. *Equilibrium Concepts in Natural Water Systems*, edited by R. F. Gould; ACS publications, 1967, 121.
(2)   Valsaraj, K. T. *Sep. Sci. Technol.* **1989**, 24, 1191.

(3)   Jaffé, P. R.; Smith, J. A.; Park, J.-W. *Proceedings of the 5th International Symposium on Interactions between Sediments and Water* ;Uppsala, Sweden, 1990.

(4)   Smith, J. A.; Jaffé, P. R.; Chiou, C. T. *Environ. Sci. Technol.* **1990**, 24, 1167.

(5)   Holsen, T. M.; Taylor, E. R.; Seo, Y.-C.; Anderson, P. R.; *Environ. Sci. Technol.* **1991**, 25, 1585.

(6)   Park, J.-W.; Jaffé, P. R.; *Environmental Engineering: Proceedings of the 1991 Specialty Conference,* sponsored by ASCE Environmental Division, edited by P. A. Krenkel, 1991, 248.

(7)   Shinoda, K.; Hutchinson, E. *J. Phys. Chem.* **1962**, 66, 577.

(8)   Kile, D. E.; Chiou, C. T. *Environ. Sci. Technol.* **1989**, 23, 832.

(9)   Tokiwa, F. *J. Phys. Chem.* **1968**, 72, 1214.

(10)   Moroi, Y.; Sato, K.; Noma, H.; Matuura, R. *J. Phys. Chem.* **1982**, 86, 2463.

(11)   Moroi, Y.; Noma, H.; Matuura, R. *J. Phys. Chem.* **1983**, 87, 872.

(12)   Smith, J. A.; Jaffé, P. R. *Environ. Sci. Technol.* **1991**, 25, 2054.

(13)   Smith, J. A.; Jaffé, P. R. *Water, Air, and Soil Pollution* **1993** (submitted).

RECEIVED January 14, 1994

Chapter 12

Kinetic Modeling of Trichloroethylene and Carbon Tetrachloride Removal from Water by Electron-Beam Irradiation

L. A. Rosocha, D. A. Secker, and J. D. Smith

University of California, Los Alamos National Laboratory, P.O. Box 1663, MS J564, Los Alamos, NM 87545

High energy electron-beam irradiation is a promising technology for removing hazardous organic contaminants in water. To explore the effectiveness of particular accelerator systems, we have formulated a simple chemical kinetics model for removing TCE (trichloroethylene) and CCl_4 in aqueous solutions. The production, recombination, and reaction of free radicals has been examined for various parameters (dose rate, pulse duration, and pulse repetition rate). Simulations show that low pulse intensities are more efficient than higher intensities in producing radicals because radical-radical recombination is dominant at higher dose rates. However, a train of short, high-dose, repetitive pulses (e.g., 100 ns, 10 kHz) can approach the removal efficiency of a continuous dose profile. Consequently, repetitively pulsed accelerators should be considered for future applications because of other specific machine advantages.

Increased sensitivity to environmental issues and the promulgation of new regulations have increased demands to reduce the amount of hazardous chemicals released to the environment through liquid waste streams and related sources. Present methods for treating aqueous-based hazardous organic wastes are constrained in scope and utility because their applicability, effectiveness, and costs are highly sensitive to the particular targeted species, water chemistry, and water quality, including solids content. In contrast to present technologies, through both laboratory and pilot-plant studies, high energy electron-beam (e-beam) irradiation has been shown to be effective and economical for the removal of hazardous organic contaminants in water (1, 2).

Conventional electrostatic electron accelerator equipment has generally been employed for high average power irradiation applications, while single-pulse accelerators have been utilized for high dose rate research. Recent technology developments have lead to a new generation of pulsed linear induction accelerators driven by solid-state electrical power conditioning elements. These are considered to be less expensive per unit delivered e-beam dose, physically smaller, modular, and more reliable than conventional electrostatic accelerators (3, 4, 5). It is speculated that these repetitively pulsed accelerators will produce better chemical destruction as well, although this remains to demonstrated. At present, we have found no data

0097–6156/94/0554–0184$08.00/0

comparing waste destruction by repetitively pulsed accelerators with that of conventional electrostatic accelerators.

Resolving questions about how one chooses a particular accelerator system architecture for overall maximum effectiveness requires an understanding of the basic removal processes and a combination of experimental testing and kinetic modeling. At Los Alamos, we have set up an e-beam laboratory to study the irradiation process using short (~65 ns), high current pulses (6). To better understand the waste removal process and e-beam machine scaling, we have also employed a computer-based chemical kinetics model to relate destruction effectiveness to electron-beam dose profiles and electron-beam machine parameters, and to make comparisons with experimental data. In this paper, we will discuss the water-radiolysis model, simple models for TCE (C_2HCl_3) and CCl_4 removal, and the experimental apparatus.

A pure-water model is being used to investigate the behavior of free radicals in irradiated solutions. We are looking at the production and recombination of these transient species as a function of dose parameters. Our goal is to maximize the average free radical concentrations over a given period of time, for a given dose, by varying dose rate, pulse duration, and pulse repetition rate, thus providing greatest destruction potential. Our preliminary simulation results show that low dose rates have advantages over higher dose rates for the efficient production of radicals. This is apparently due to nonlinearities within the water model that favor radical recombination over radical production at higher dose rates.

With TCE or other pollutants present, it has been postulated (although not demonstrated) that the formation of radical adducts and their subsequent reactions will produce favorable nonlinear effects that possibly make the pulsed case more advantageous in terms of chemical efficiency. We will eventually explore this mechanism through more complete models.

Computer Simulation of Kinetics

The process of e-beam irradiation is best understood in aqueous solutions, in which sizable quantities of the free radicals e^-_{aq}, H, and OH, as well as the more stable oxidant H_2O_2, are produced. These highly reactive species react with organic contaminants to produce CO_2, H_2O, salts, and other compounds which are no longer hazardous. Our simulation effort is divided into two parts: a pure-water model and a simple model for TCE and CCl_2 removal.

The chemical kinetics code that we employ is zero-dimensional and therefore assumes that the electron-beam energy is uniformly deposited in space. From accepted track theories of water radiolysis, we know that e-beam energy deposition is not uniform; however, this does not necessarily imply that the code will not have predictive capability once benchmarked against experiments. The computer code, called MAKSIM (7), will accommodate 200 chemical reactions and 60 different reacting species. Because in most cases only a few of the species actually react with each other, sparse matrix inversion techniques are utilized by the code. Also, because such chemical systems usually involve a wide range of reaction rates, stiff differential-equation-solving techniques must be employed. MAKSIM uses the Gear (8) integration algorithm to handle a stiff equation set.

Pure-Water Model

The predominant radical reactions that take place within the pure-water model are listed in Table I below. The radicals are created by the first extremely fast reaction and the radical concentrations are determined by known yields. These yields

(G-values) for radical production by beam electrons are 2.7, 2.7, 0.55, and 0.71 molec/100eV for e^-_{aq}, OH, H, and H_2O_2, respectively.

Table I. Predominant Radical Production and Loss Reactions

Reaction	Rate Constant $(L \cdot mol^{-1} \cdot s^{-1})$
$e + H_2O \rightarrow e^-_{aq} + OH + H + H_2O_2 + H^+$	(extremely rapid)
$e^-_{aq} + H_2O_2 \rightarrow OH + OH^-$	1.2×10^{10}
$e^-_{aq} + H^+ \rightarrow H$	2.2×10^{10}
$e^-_{aq} + OH \rightarrow OH^-$	3.0×10^{10}
$H + OH \rightarrow H_2O$	2.0×10^{10}
$OH + OH \rightarrow H_2O_2$	5.5×10^9
$e^-_{aq} + H \rightarrow H_2 + OH^-$	2.5×10^{10}
$e^-_{aq} + e^-_{aq} \rightarrow OH^- + OH^- + H_2$	5.0×10^9

This pure water model is based upon reactions and rate constants described in similar models found in the literature (9). The referenced paper also contains experimental data that supports the validity of the model quite well. In our calculations, we are interested in relating radical production and recombination to the dose profile and, therefore, comparing different types of accelerators. We are particularly interested in radical recombination since the primary production of radicals is assumed to be linear with dose rate (determined by the G-value), and the secondary production of radicals from the water reactions appears to be negligible. Note in Table I that two of the recombination reactions are bi-linear in nature, while the remaining loss reactions are nonlinear because, in each case, both reactants are produced linearly with dose.

To examine the effect of dose rate on the efficiency of radical production (yield), we have applied the pure-water model to a fixed duration e-beam pulse (65 ns) and varied the total dose applied during the pulse from 1 to 8 Mrad. In this way, the average dose rate is varied from about 15×10^{12} rad/s to 123×10^{12} rad/s. In this set of calculations, the radical yields are described by the equation

$$G_R(t) = \frac{100 \, N_A [R(t)]}{\rho \cdot D(t)} \quad , \qquad (1)$$

where $G_R(t)$ is the yield of a particular radical (radicals per 100eV of deposited e-beam energy) as a function of time t, N_A is Avogadro's number, ρ is the density of water, $[R(t)]$ is the radical concentration in moles/L, as a function of time t, and $D(t)$ is the dose delivered as a function of time t, in eV/kg of water. Figures 1 to 3 show plots of the calculated $G_R(t)$ versus time t.

Two significant results were obtained from these calculations: for the lower dose rate (1-Mrad total dose), the radical yields fall to less than 50% of their initial values by the end of the pulse, which is caused solely by the rate of radical recombination; for the higher dose rate case (8-Mrad total dose) the yields fall even further (to about 10% of their initial values), which indicates that the lower dose rate is

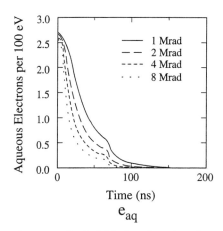

e_{aq}

Figure 1. Aqueous electron radical yield plotted vs time for a single 65-ns pulse.

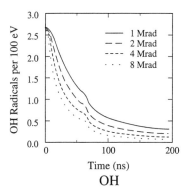

OH

Figure 2. Hydroxyl radical yield plotted vs time for a single 65-ns pulse.

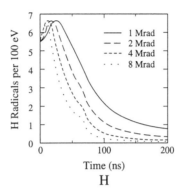

Figure 3. Hydrogen radical yield plotted vs time for a single 65-ns pulse.

more efficient for radical production and subsequent waste destruction. The radical yields decrease at a faster rate as the dose rate increases, which reflects the presence of the nonlinear radical recombination processes. If there were no recombination effects, the radical yields would not change with dose rate. Note that for all three radical species, the yields drop off most rapidly at 65 ns; the time at which there is a total loss of excitation to the system by the e-beam. In the case of the hydrogen radical, the G value rises slightly during the first half of the irradiation pulse. This is caused by the production of H in the water reactions, which are scavenging the e_{aq} and OH radicals and producing H. Lastly and importantly, these calculations show that the rate of radical decay is much slower for the H and OH radicals than the e^-_{aq} radicals, which possibly makes them more effective for waste destruction.

Figures 4 and 5 show running averages of radical yields for a DC dose profile and a 10-kHz repetition rate pulsed dose of duration 100 ns, respectively. In these simulations, the average dose rate was held constant at 1 Mrad/s, which increases the peak dose for the 10-kHz case. The running averages are given by the following equation:

$$[\hat{R}(\tau)] = (1/\tau) \int_0^\tau [R(t)]dt \quad , \tag{2}$$

where $[\hat{R}(\tau)]$ is the concentration of a particular radical in moles/L averaged over time τ and $[R(t)]$ is the radical concentration in moles/L, as a function of time t.

From the figures, it can be seen that the 10-kHz pulsed profile produces average radical concentrations almost as large as those for the DC dose profile. This implies that suitably applied repetitively pulsed dose profiles can produce removal efficiencies similar to those of a DC dose profile.

It should also be noted that in the case of DC dose, the average e^-_{aq}, H, and OH concentrations rise at a constant rate at times less than a few tenths of 1 ms, and drop off thereafter. Therefore, because the dose is also increasing at a constant rate, the yields at times less than a few tenths of 1 ms are constant. After 1 ms, the yields of all three species drop significantly. This might imply that a residence time or pulse duration of a few tenths of 1 ms is optimum for destroying waste when a dose rate of 1 Mrad/s is used. The trend of average e^-_{aq}, H, and OH concentrations falling off before the time of 1 ms is also apparent in the 10-kHz case.

TCE and CCl$_4$ Removal Models

TCE and CCl$_4$ removal are being modeled by one-step reactions with the aqueous electron, the hydroxyl radical, and the hydrogen radical:

$$e^-_{aq} + TCE, CCl_4 \rightarrow products,$$

$$OH + TCE \rightarrow products,$$

$$H + CCl_4 \rightarrow products.$$

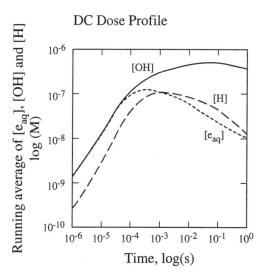

Figure 4. Plot of running average of aqueous electron, hydroxyl, and hydrogen radical concentrations vs time for a DC dose profile.

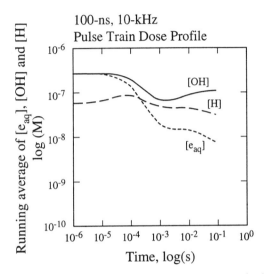

Figure 5. Plot of running average of aqueous electron, hydroxyl, and hydrogen radical concentrations vs time for a 100-ns, 10-kHz repetitive pulse train dose profile.

These reactions simplify the destruction mechanisms of TCE and CCl$_4$ that involve more than a dozen reactions for each case. Reaction products such as the Cl• are not incorporated into the models, and therefore, one would expect to obtain better destruction efficiencies with the simplified models due to the absence of radical scavenging by reaction products (e.g., aldehydes, formic acid). Simulation parameters are chosen to match our accelerator's dose profile as well as the initial TCE and CCl$_4$ concentrations used in our destruction experiments, and those at the Florida International University - University of Miami (FIU-UM) facility (1).

In Figure 6, the fractional destruction of TCE and CCl$_4$ are plotted as functions of initial concentration and dose. The dose is applied over a constant 65-ns pulse duration, which simulates our LANL accelerator dose profile. As expected, as more chemical is removed, more dose is required to achieve greater and greater fractional removal. Also, at low doses, for TCE, the fractional destruction depends greatly upon initial concentration; but as the dose increases, this dependency decreases to almost zero. For CCl$_4$ the dependency on initial concentration is much less pronounced.

Figures 7 and 8 show the destruction of TCE and CCl$_4$ versus time for a 4-Mrad dose applied over 65 ns. Note how the e$^-_{aq}$ concentrations die away almost immediately, while the H and OH radicals have a greater lifetime, and therefore can destroy waste for a greater period of time. Ninety-two per cent of the TCE, and 70% of the CCl$_4$ is eventually destroyed. The 92% value agrees fairly well with our measured value of 87% in pulsed-accelerator experiments at Los Alamos.

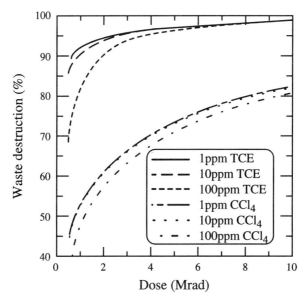

Figure 6. Fractional destructions of TCE and CCl₄ plotted vs applied dose for a family of different initial concentrations.

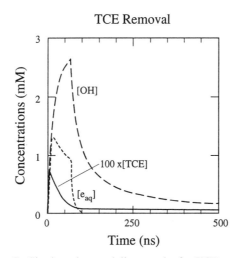

Figure 7. Single-pulse modeling results for TCE removal.

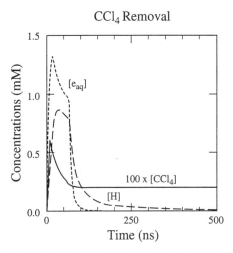

Figure 8. Single-pulse modeling results for CCl₄ removal.

Tables II and III show TCE and CCl₄ destruction vs dose and method of application of the dose. The initial concentrations and applied doses are chosen in an attempt to reproduce results from the FIU-UM facility (*1*). Four cases are examined for both compounds:

- Single 100-ns pulse
- 100-ns, 1-kHz repetition rate pulse train
- 100-ns, 10-kHz repetition rate pulse train
- DC irradiation.

Table II. TCE Destruction Calculated for Different Doses and Methods of Application

Dose Method	Fractional Destruction (%)	
	100-krad dose	**150-krad dose**
100-ns pulse	37.5	46.1
1-kHz pulse train	68.0	92.2
10-kHz pulse train	68.8	95.9
DC	69.4	96.8

Notes: initial TCE concentration is 100 ppm; residence time is 0.1 sec.

Table III. CCl₄ Destruction Calculated for Different Doses and Methods of Application

Dose Method	Fractional Destruction (%)	
	50-krad dose	100-krad dose
100-ns pulse	29.3	34.1
1-kHz pulse train	77.9	91.2
10-kHz pulse train	81.2	94.1
DC	81.8	94.9

Notes: initial CCl₄ concentration is 10 ppm; residence time is 0.1 sec.

The following results are derived from the tables: (1) both the 1-kHz and 10-kHz pulse trains approach the DC case in destruction effectiveness; (2) a single pulse is not nearly as effective as the DC or repetitively pulsed cases (this is also supported by Figures 4 and 5); (3) to reproduce the results from the FIU-UM facility, our initial concentrations had to be increased by a factor of 50.

Potable water or natural water normally contains radical scavengers (e.g., carbonates) that decrease the average radical concentrations and consequently decrease the destruction efficiency. Because our model presently does not account for radical scavengers, the required increase in concentration in our calculations is in the right direction. Also, our model does not include the whole destruction mechanisms for TCE and CCl₄. These mechanisms involve some scavenging that would further decrease radical populations. It is impossible to determine the extent of this scavenging because most of the reaction rates are unknown at present. Future work will be directed at refining the model.

Experimental Apparatus

We are currently testing a high-dose/high-dose rate electron accelerator for the irradiation of aqueous-based organics. Table IV lists the accelerator specifications and Figure 9 shows an equipment schematic diagram. The accelerator consists of three major parts: a high-voltage pulse generator (Marx generator), an electrical pulse-forming line (triaxial oil-insulated Blumlein transmission line), and electron gun (cold-cathode relativistic electron-beam diode). A reaction chamber, which contains the aqueous-based organic sample to be treated, is attached to the output section of the electron gun. Diagnostic ports and windows are also provided for chemical sampling and laser-based diagnostics. This accelerator differs from existing commercial continuous-duty electron accelerators in that its dose is pulsed rather than continuous. The pulsed cold-cathode electron accelerator is well suited for a research application because it permits separate transient measurements of the production of free radicals and the subsequent destruction of organics in the water and provides a high dose. Using this setup, we envision routine delivery of electron-beam doses of 1 to 10 Mrads in water samples. Additionally, the development of analytical techniques for real-time, *in situ* identification and quantification of reactants, products, reaction rates, and e-beam dose are extremely desirable. In our experiments, we are initially concentrating on measuring the aqueous electron concentration because of its importance in understanding the waste removal process and in providing time-dependent data to verify the pure-water and TCE models. The aqueous electron absorption spectrum is well known and, as shown in Figure 9, its measurement is accomplished with the red laser beam (near 700 nm).

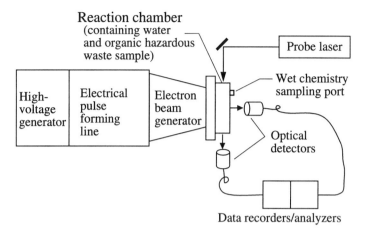

Figure 9. Schematic diagram of experimental equipment setup, including accelerator, sample chamber, and laser-based diagnostics.

Table IV.   Specifications for Los Alamos High-Dose E-beam Facility

Parameter	Value
Electron energy (voltage)	1 - 4 MeV
E-beam current density	2 kA/cm^2
E-beam cross section	3 x 10 cm^2
Electrical pulse duration	60 - 70 ns
Electron penetration depth	0.5 - 1.0 cm
Transport efficiency for e-beam	~50%
Electron dose in water samples	~1 - 10 Mrad

Future Electron-Beam Systems

Although the computer modeling shows that a continuously applied dosage is more efficient in destroying waste than the same amount of pulsed dosage, the modeling does show that a repetitively pulsed machine can approach the efficiency of a DC machine when pulsed at high pulsed repetition rates (e.g., 10 kHz). Consequently, repetitively pulsed accelerators should be considered for future applications involving the destruction of hazardous organic wastes by e-beams.

The advantages of DC electrostatic accelerators are mature technology, simple architecture for moderate voltages (1 to 3 MV), demonstrated industrial service, and reasonable power efficiency. The advantages of RF electrostatic accelerators are mature technology, relatively simple architecture, a compact and efficient transformer, demonstrated industrial service, and reasonable power efficiency. The disadvantages of DC and RF electrostatic accelerators are limited current (due thermionic-emission cathodes), full, high-voltage hold-off insulation requirements, a considerable parts count in the high-voltage section, production of exit foil window hot spots with a scanned beam, and scanner duty-factor losses.

In comparison, the advantages of pulsed linear induction accelerators are mature technology, simpler and less stringent high voltage isolation, ease of scaling to high power and high voltage (due to its modularity), smaller size (large, high-voltage transformer not required), and wide power control using variable pulse frequency. The main disadvantages of this technology are that it has not fully been demonstrated for industrial service, and requires more knowledgeable servicers.

The advantages of pulsed cold cathode accelerators are mature technology for single-shot, very high peak powers and doses for research applications, simple high-voltage hold-off, and scalability to large beam areas (square meters). This style of accelerator is best used for research purposes because it is not well suited for rep-rate service.

Conclusions

A pure-water model was used to investigate the behavior of free radicals in irradiated solutions by calculating the production and recombination of free radicals as a function of dose parameters. Our goal was to provide the greatest destruction potential by maximizing the free radical concentrations and yields for a given dose by varying dose rate, pulse duration, and pulse repetition rate. Our preliminary simulation results show that low dose rates have advantages over higher dose rates for the efficient production of radicals. This is apparently due to nonlinearities within the water model that favor radical recombination over radical production at higher dose rates.

Also, we have found that a suitable application of a train of repetitive short-duration pulses (e.g., 100 ns) gives radical concentrations and fractional removals similar to a DC applied dose.

Additionally, we have discussed the relative power-conditioning and architecture advantages that pulsed linear induction accelerators hold compared to conventional electrostatic electron accelerators. These are primarily high-voltage isolation, ease of scaling to high power and high voltage, modularity, smaller size (large, high-voltage transformer not required), and wide dynamic range of power control using a variable pulse frequency.

Literature Cited

1. Cooper, W. J.; Nickelsen, M. G.; Waite, T. D.; and Kurucz, C. N., "High Energy Electron Beam Irradiation: an Innovative Treatment Process for the Treatment of Aqueous Based Organic Hazardous Wastes," *Fifth Annual Aerospace Hazardous Waste Minimization Conference*, Costa Mesa, California, **May 1990**.
2. Kuruz, C. N., Waite, T. D., and Cooper, W. J., "Full-Scale Electron Beam Treatment of Hazardous Wastes - Effectiveness and Costs," *45th Purdue Industrial Waste Conference Proceedings*, **May 1990**, p. 539.
3. Harjes, H. C.; Penn, K. J.; Reed, K. W.; McClenahan, C. R.; Laderach, G. B.; Wavrik, R. W.; Adcock, J.; Butler, M.; Mann, G. A.; Martinez, L.; Morgan, F. A.; Weber, G. J.; and Neau, E. L., "Status of the Repetitive High Energy Pulsed Power Project," *Proceedings of the 8th IEEE Pulsed Power Conference*, **June 1991**.
4. Jacob, J. H., "Reliable Low Cost Induction Accelerator System for Treatment of Industrial and Municipal Wastewater," *private communication*, Science Research Laboratory, Sommerville, Massachusetts, **Aug. 1991**.
5. Rosocha, L. A.; Dyer, R. B.; Holland, R. F.; and Wampler, F. B., "Electron-Beam Sources and Diagnostics for the Treatment of Aqueous-Based Hazardous

Organic Compounds," presented at *AIChE Annual Meeting*, Los Angeles, California, **Nov. 1991**.

6. Rosocha, L. A.; Allen, G. R.; Coogan, J. J.; Kang, M.; Smith, J. D.; McCulla, W. H; Buelow, S. J.; Dyer, R. B.; Anderson, G. K.; Wampler, F. B.; Tennant, R. A.; and Wantuck, P. J., "Advanced Chemical Processes for Hazardous Waste Destruction Study," *Chemical and Laser Sciences Division Annual Report 1991*, Los Alamos National Laboratory report LA-12331-PR, **June 1992**.

7. Carver, M. B.; Hanley, D. V.; and Chaplin, K. R., "MAKSIMA-CHEMIST A Program for Mass Action Kinetics Simulation Manipulation and Integration Using Stiff Techniques," Chalk River Nuclear Laboratories report AECL-6413, **1979**.

8. Gear, C. W., "Automatic Integration of Ordinary Differential Equations," *Comm. ACM* **1971**, Vol. 14, p. 176.

9. Boyd, A. W; Carver, M. B.; and Dixon, R. S., "Computed and Experimental Product Concentrations in the Radiolysis of Water," Radiat. Phys. Chem. **1980**, Vol. 15, p 177.

RECEIVED December 1, 1993

Chapter 13

Enhancement of Pentachlorophenol Biodegradation by Fenton's Reagent Partial Oxidation

Judith Bower Carberry[1] and Sang Ho Lee[2]

[1]Department of Civil and Environmental Engineering,
University of Delaware, Newark, DE 19716
[2]Star Environmental Consulting Company, Taegu, Korea

PCP, the most highly chlorinated phenol, has been widely used as a wood preservative and is often present at toxic levels at production sites and at treatment sites. Fenton's Reagent, a hydrogen peroxide solution containing ferrous iron as a catalyst to produce the strong oxidizing agent, hydroxyl radical, was used in this study to partially oxidize PCP. Partial oxidation products were then subjected to biodegradation by a selected microbial consortium from PCP contaminated soil. The parent PCP and partially oxidized PCP systems were compared with respect to levels of destruction and rates of biodegradation. PCP was chemically oxidized to a level of only 3%, and the subsequent biodegradation was enhanced to a level almost three times that of the untreated control and at a rate approximately twenty times faster. Data modeling using the Haldane Equation indicated that the Haldane Constant, a toxicity index, was reduced to a level 4 times less from the slight chemical oxidation by Fenton's Reagent.

Pentachlorophenol (PCP) is the second most widely-used pesticide in the United States. PCP and its sodium salt (Na-PCP) are biocides since they are lethal to a variety of living organisms including plants and animals. PCP is registered by the U.S. Environmental Protection Agency (1) for use as a bactericide, insecticide, fungicide, herbicide and algicide. Also a large amount of PCP is used to control algae and fungi in cooling towers at power plants and manufacturing factories. Although PCP and its derivatives have many applications, PCP is mostly used in the wood preservation industry. Worldwide production of PCP in 1983 was estimated at 50,000 metric tons, of which about 2300 metric tons was produced in the United States (2). Approximately 80% of U.S. PCP production is used for commercial wood treatment, and there are approximately 600 treatment plants in the United States (3). Accidental spillage and improper disposal of PCP at production plants and at wood-preserving facilities have resulted in extensive contamination of soil, surface water, and ground water aquifers (4).

0097–6156/94/0554–0197$09.08/0

The toxicity of chlorinated phenols tends to increase with the degree of chlorination. PCP, the most highly chlorinated phenol, is recalcitrant to degradation by microorganisms and therefore accumulates in the environment (5). PCP has been detected in human and mammal tissues, and in fishes and birds. PCP is acutely toxic to a variety of organisms and mammals as an inhibitor of oxidative phosphorylation (6-8), and PCP is a well known fat-soluble chemical accumulating in fish directly through the skin or through the food chain (9). Thompson and Kaiser (5) examined the toxicity effect of several chlorophenols, including PCP, with *Bacillus subtilis* isolated from activated sludge. They demonstrated that the more highly chlorinated the phenol, the greater the toxicity to microorganisms.

PCP is presumed to be very resistant to microbial degradation due to its highly chlorinated organic nature; the feasibility of PCP biological treatment, however, has been the subject of several research papers (10-11) in which a number of researchers studied the biodegradation of PCP with a variety of microorganisms. PCP biodegradation in soils was examined by several researcher (12-17) and in aquatic environments by Melcer and Bedford (11) and Godsy (3). These studies were supplemented by studies using consortia (18) and pure cultures of bacteria (19, 20, 7). Further evidence of PCP biodegradation was provided by documenting the production of inorganic chloride ions (21, 22) and $^{14}CO_2$ formation from radiolabeled PCP (23-25, 7). Later, Volskay and Grady (26) tested the toxicity of 33 priority pollutants to activated sludge and found PCP to be the most toxic and resistant of those toxic chemicals tested.

Crawford and Mohn (27) reported that within one week a PCP-degrading *Flavobacterium* removed less than 100 ppm PCP from a soil sample containing 10 - 20% water but could not effectively remove PCP at a concentration of 500 ppm. Steiert and Crawford (28) reported that microorganisms degrading PCP remove at least one chlorine atom prior to ring cleavage since, in general, a variety of chlorophenols must be di-hydroxylated by oxygenases for aerobic microbial catabolism. Halogen atom substituents on the aromatic ring hinder the electrophilic attack on the ring by withdrawing electrons from the ring. Rochkind-Dubinsky *et al.* (29) corroborated the same mechanism and sequence of these degradation reactions using *P. putida.*

Enhanced levels of xenobiotic biodegradations have been observed in the presence of mixed substrates, and, often, co-substrates are required to support the growth of degrading microorganisms. The co-substrates may take part directly in the catabolism of xenobiotic compounds by supplying co-factors for the enzyme pathway or supplying energy for the transport of the xenobiotic compounds (30). Therefore, the simultaneous utilization of mixed substrates results in enhanced removal efficiencies, compared to the removal efficiency of an individual xenobiotic compound as a sole carbon and energy source. Enhanced removal of mixed substrates generally produces a higher level of biomass concentration.

Alexander (31) defined the co-metabolism mechanism as "the fortuitous degradation of one compound by an enzyme which routinely acts on primary substrate." Overall biodegradation of halogenated compounds can be enhanced by supplemental addition of co-substrate (32-34). Co-metabolism studies of PCP uptake have been conducted by Klecka and Maier, (35), Thompson and Strachen (36), and Hendricksen *et al.* (37). Results of these studies demonstrated both an increase in PCP uptake and an increase in microbial growth.

Since regulatory agencies would be reluctant to approve the addition of co-substrates to a contaminated waste site, this study was undertaken to determine biodegradation enhancement effects and mechanisms from the generation of co-substrates due to oxidation of PCP by Fenton's Reagent.

Fenton proposed that hydrogen peroxide and ferrous salts, known as "Fenton's Reagent," could produce the powerful inorganic oxidant, hydroxyl radical (*38*). Later, Haber and Weiss (*39*) demonstrated that hydroxyl radical was a strong oxidant in their reaction systems containing a wide variety of organic compounds. They proposed the following reaction mechanism of ferrous ion and hydrogen peroxide reaction:

$$Fe^{2+} + H_2O_2 \longrightarrow Fe^{3+} + \cdot OH + OH^- \tag{1}$$

where $\cdot OH$ is the hydroxyl radical, a very strong oxidizing agent. In the presence of organic compounds, the hydroxyl radical, $\cdot OH$, may react via mechanisms involving addition of $\cdot OH$ or extraction of a hydrogen atom, as shown in the following reactions;

$$\cdot OH + R \longrightarrow \cdot ROH \tag{2}$$

$$\cdot OH + RH_2 \longrightarrow \cdot RH + H_2O \tag{3}$$

In both reactions, organic free radicals react further with Fe^{3+}, oxygen, hydrogen peroxide, $\cdot OH$ or with other organic radicals (*40*).

Schumb (*41*) demonstrated the reaction of Fenton's Reagent with substituted halophenols and found that the greater the degree of halogen substitution, the slower the reaction rate. With one mole of ferrous ammonium sulfate per mole of phenol and various concentrations of hydrogen peroxide at 50 °C and pH of 4, the initial concentration of phenol decreased quickly within two minutes of initial reaction time and then remained unchanged thereafter, presumably because the total amount of hydrogen peroxide had been consumed.

Barbeni *et al.* (*42*) examined the chemical oxidation of chlorophenols (monochlorophenol, dichlorophenol, and trichlorophenols) using Fenton's Reagent in aqueous solution. They found that the oxidation rate of 3,4-dichlorophenol increased with increasing ferrous ion concentration at a fixed concentration of H_2O_2. In addition, they illustrated the efficiency of phenol oxidation at high concentrations, such as the 1000 ppm concentration in wastes produced by the pulp and paper industry. Watts and his co-workers (*43*) investigated the Fenton's Reagent oxidation of pentachlorophenol in contaminated soil. They demonstrated that the chemical oxidation of PCP released chloride ions, and that Fenton's Reagent oxidation was maximized due to enhanced hydroxyl radical generation in a pH range of 2 to 4.

These previous studies demonstrated the efficacy of using Fenton's Reagent for its ability to generate the strong oxidizing agent, hydroxyl radical, to treat high concentrations of PCP. The complete chemical oxidation of PCP, however, is very expensive, and its increased demand in soils contaminated with PCP has been demonstrated. The purpose of this study, therefore, was to demonstrate the benefit of reducing toxic PCP concentrations and generating co-substrates by partial chemical oxidation in order to enhance PCP removal by a selected microbial consortium (SMC) and increase its growth.

Methods and Analyses

Reagent grade PCP (99 + % purity; Fluka AG, Fluka, Switzerland) was used as a sole carbon source for the PCP-degrading microorganisms and activated sludge. A stock solution of PCP was prepared at an elevated pH of 11 with 1 N NaOH to increase the solubility of PCP in the aqueous phase.

For the partial chemical oxidation pretreatment by Fenton's Reagent, the ferrous sulfate solution ($FeSO_4 \cdot 7H_2O$, Fisher Scientific Co., Fair Lawn, NJ) was prepared at a concentration of 1 g/L as Fe. Hydrogen peroxide as 35 % H_2O_2 was purchased from Fisher Scientific Co. Solutions of 1, 5 and 10 Mm H_2O_2 in distilled water were prepared for chemical oxidation. The ferrous ion concentration was in excess of the optimum ferrous ion versus hydrogen peroxide ratio determined by Barbeni *et al.* (*42*). Three different initial hydrogen peroxide concentrations were added to each set of triplicate reactor vessels at the following molar ratios of H_2O_2 to PCP: 0.1:1, 0.5:1 and 1:1, designated as FR(0.1:1), FR(0.5:1), and FR(1.0:1). A stepwise 25 ml aliquot of hydrogen peroxide solution was added into each reactor at a pH of 3.5 and at 20 ± 1 °C as determined by Bishop, *et al.* (*44*), and at one-hour intervals over a four hour period as described by Bowers *et al.* (*45*).

A selected microbial consortium (SMC) of PCP-degrading microorganisms was prepared and maintained as previously described by Lee and Carberry (*46*). The two microbial strains isolated and identified by the Hewlett Packard model 5898A MIS System following long-term culture maintenance were *Pseudomonas putida* and *Pseudomonas aeruginosa*.

All biodegradation experiments with or without partial oxidation pretreatment were carried out using an electrolytic respirometer system (Bioscience Management, Inc., Bethlehem, PA) to provide oxygen and to determine accumulated oxygen uptake in each 1 liter reactor. The system consisted of a main control module (Model ER-101) and a water bath controlled at a temperature of 20 °C mounted on a gang magnetic stirrer so that the 1 liter flasks contained therein could be mixed continuously.

Biodegradation of 1mM PCP solutions, untreated by Fenton's Reagent, were designated CONTROL for the SMC microbial system. For biodegradation experiments following chemical oxidation, the solutions in the reactors were adjusted to a pH of 7 to 7.2 with 1 N NaOH or 2 N H_2SO_4, any ferric hydroxide precipitate carefully removed by decantation, and a uniform 120 mg/L titer of microorganisms introduced into each reactor for the subsequent biodegradation. The subsequent biodegradation rates were measured for 12 days by monitoring differences in several parameter concentrations, as described below. These experiments were designated as FRSMC, according to the dose of Fenton's Reagent used.

For PCP analysis, a high performance liquid chromatograph was used during both the chemical oxidation and biodegradation processes. A Varian HPLC (Varian Instrument Group, Walnut Creek, CA) system was used, consisting of dual pumps to mix mobile phases, a Model 2550 UV detector at a wave length of 280 nm, and a Varian model 4290 Integrator. Following the partial chemical oxidation reactions, samples were adjusted to Ph 11 to optimize PCP solubility and filtered through 0.22 μm membrane filters (Millipore, Bedford, MA) before each injection. Samples removed from the CONTROL or FRSMC systems required no pH adjustment. All aqueous filtrates were injected using a Rheodyne 7175 a 20 μl injector loop onto a reverse-phase column (a 250 mm Nucleosil C-18 packed column). The isocratic elution was pumped through the column at a rate of 1 ml/min and was composed of 88% methanol with 1% acetic acid and of 12% deionized water with 1% acetic acid. The calibration curve for several PCP concentrations versus integrator peak area was characterized by an R^2 factor, the coefficient of determination, equal to 0.999. For each PCP experiment, the HPLC instrument was calibrated each sampling day, and samples were prepared for analysis and injected immediately. Statistical description of HPLC and observed values precision are presented in Table 1. These results were determined by measuring a series of 15 PCP injections into the HPLC, and they illustrate the extent of scatter. The standard deviations of PCP concentrations were also calculated with respect to the average and shown in Table 1.

Table 1. Statistical Analysis of HPLC Precision of Observed PCP Concentration

nominal concentration (mg/L)	100.00	300.00
mean from 15 observations (mg/L)	99.7	299.37
standard deviation	0.6	1.87

The potassium iodide-sodium thiosulfate titration method was used to determine residual concentrations of hydrogen peroxide in each chemical oxidation system (*47*). Also a peroxide test strip (EM Science, Gibbstown, NJ) was used to ensure the total disappearance of hydrogen peroxide before the application of microorganisms.

Inorganic chloride ion concentrations were determined with a Fisher 825 MP digital Ph meter equipped with an Orion model 94-17B chloride electrode (*25, 22*) for both the chemical oxidation and biodegradation processes. Inorganic chloride ion concentration was determined using a calibration curve plotted from the molarity of a series of KCl standards. The R^2 factor from this determination was 0.999.

Each concentration of PCP-degrading microorganisms was determined by optical density measurements using a spectrophotometer (Spectronic 21, Bausch & Lomb Inc., Rochester, NY). For the calibration curve of selected PCP-degrading microorganisms, 50 ml of each suspension was diluted and filtered through a tared 0.22 μm membrane filter (Millipore, Bedford, MA). The dry weight of microorganisms at different dilutions was measured according to the procedure described by Karns *et al.* (*33*). Optical density readings and the dry weight of microorganisms were correlated at each dilution to develop the calibration formulation. Since these calibration curves were developed by sequential dilution of this microbial suspension, any error between the microbial mass and microbial viability was proportional to the microbial concentration. The R^2 factor for this calibration was 0.999.

Oxidation products from the PCP chemical oxidation were analyzed as acetylated derivatives by GC/MS. Those acetylated compounds were identified by GC/MS according to the following modified procedure developed by Haggblom *et al.* (*48*). A 7 ml aqueous sample was acetylated by adding 0.7 ml of 1 M K_2CO_3 and 0.2 ml of acetic anhydride. The acetylated compounds were extracted using a Mixxor (Genex Corporation, Gaitherburg, MD) two times with 7 ml of HPLC-grade n-pentane; the extract was then concentrated to 100 μl and analyzed by GC/MS (HP 5890 Hewlett Packard Co., Palo Alto, CA) using a HP-1 (Hewlett Packard) capillary column with chromatograph grade helium carrier gas and a mass selective detector (HP 5970 Hewlett Packard Co., Palo Alto, CA). The temperature program consisted of 80 °C for 1 minute, followed by a temperature gradient period at 8 °C/min up to 260 °C.

The specific substrate uptake rates at each initial substrate concentration were determined by integrating the left hand side of Equation 4 and plotting S versus Xt.

$$\frac{1}{X}\frac{dS}{dt} = -K \tag{4}$$

The results were then equated to the Michaelis-Menten function, shown on the right hand side of Equation 5.

$$\frac{1}{X}\frac{dS}{dt} = \frac{kS}{K_m+S} \tag{5}$$

where, $(1/X)(dS/dt)$=specific substrate uptake rate, (time^{-1})
X=average biomass concentration during biological reaction
$\quad X=(X_o + X_t)/2$, X_o, X_t; the biomass concentration
$\quad$ at the time t=0 and t=t, respectively (mass)/(volume)
k=maximum substrate uptake rate constant, (time^{-1})
K_m=the Michaelis-Menten constant; the substrate concentration at
$\quad$ which the specific uptake rate is half the maximum rate,
$\quad$ (mass substrate)/(volume)
S=substrate concentration (mass substrate)/(volume)

The characteristic Michaelis-Menten constants were determined using a curve fitting software program which employed a cubic spline interpolation routine and a corresponding evaluation of the best fit y-parameter for the Michaelis-Menten function. The routine used a trial and error fitting procedure until the resulting error reached a minimum and, then, values for the two constants, k and K_m were provided. These results were subjected to a t-test to determine whether the resulting constants characteristic of each environmental condition were the same or significantly different at a probability level equal to 0.05 using a Sigmaplot^R software program (Jandel Scientific, Corte Madera, CA).

Specific microbial growth rate was similarly evaluated for experimental growth phases by integrating Equation 6 and plotting ln X/X_o vs t.

$$\frac{1}{X}\frac{dX}{dt} = \mu \qquad (6)$$

The specific growth rate of microorganisms on non-inhibitory substrates is given by the Monod Equation.

$$\mu = \frac{\mu_{max} S}{K_s + S} \qquad (7)$$

where, μ_{max} is the maximum specific growth rate constant and K_s is the substrate concentration at which the specific growth rate is the half the maximum rate. For toxic or inhibitory organic chemicals, the Haldane equation can be utilized to demonstrate the relationship between S and μ. An increase in the initial concentration of toxic organic chemical results in higher specific growth rates until a peak growth rate, μ^*, at the critical substrate concentration, S^* is observed. Any increase in the substrate concentration above S^* results in a decreasing growth rate. The Haldane function is presented in Equation 8.

$$\mu = \frac{\mu_{max} S}{S + K_s + \dfrac{S^2}{K_i}} \qquad (8)$$

K_i is the Haldane inhibitory constant which describes the depression of growth rate due to substrate inhibition. As inhibition is decreased, K_i increases and the formula approaches the Monod Equation. When inhibitory concentrations of PCP are introduced to microorganisms, a much longer lag time is observed prior to the onset of exponential growth. The lower the value of K_i, the longer the time required for the onset of exponential growth. While K_s characterizes the slope of the ascending leg of the specific PCP uptake rate with increasing PCP concentration curve, K_i characterizes the slope of the descending leg of the curve at PCP concentrations greater than S^*.

The critical substrate concentration, S^*, can be calculated by taking the derivative of Equation 8 with respect to S, setting the derivative equal to zero, and solving for S^*. This result is given by Equation 9.

$$S^* = \sqrt{K_s K_i} \qquad (9)$$

The peak growth rate, μ^* can be determined by substituting Equation 9 into Equation 8, yielding Equation 10.

$$\mu^* = \frac{\mu_{max}}{1 + 2\sqrt{\dfrac{K_s}{K_i}}} \qquad (10)$$

The biokinetic constants of the Haldane equation, therefore, provide S^* and μ^* values to compare the CONTROL and the pretreated systems. These Haldane constants were determined using a non-linear curve fitting software program which employed an interpolation routine and corresponding evaluation of the best fit y-parameter for the Haldane equation. By using a trial and error fitting procedure until the resulting error reached a minimum, the routine returned three biokinetic constants, μ_{max}, K_S, and K_i. These results were analyzed by a t-test using a Sigmaplot^R software program (Jandel Scientific, Corte Madera, CA) to determine whether the resulting constants characteristic of each environmental condition were the same or significantly different at a probability level of 0.05. S^* and μ^* values were calculated from these constants using Equations 9 and 10, respectively.

In order to examine the co-metabolic mechanism for enhanced PCP uptake, experiments were carried out in which the PCP solutions were supplemented with the single chemical oxidation product identified by GC/MS, tetrachlorocatechol (TeCC) and, subsequently, with aliquots of oxidized PCP solution. At the beginning of the experiment (PHASE I), 5 ppm or 15 ppm of the oxidation product, tetrachlorocatechol (TeCC) was added to each reactor. SYSTEM 1 received 15 ppm of TeCC, and SYSTEM 2 received 5 ppm of TeCC. Composition of the two systems are summarized in Table 2. After 15 days, PHASE II was initiated for each system by adding 150 ml of 1mM PCP solution oxidized by Fenton's Reagent to each reactor, and after another eight days PHASE III began by adding another 100 ml of PCP solution oxidized by Fenton's Reagent. These procedures documented the effects of sequential additions of single and multiple co-substrates on biodegradation.

Table 2. PHASE I solution compositions in for enhancement mechanism study

	SYSTEM 1 (15 ppm TeCC)		SYSTEM 2 (5 ppm TeCC)	
	volume (ml)	concentration	volume (ml)	concentration
PCP solution	450	135 ppm	450	135 ppm
microorganisms	100	850 mg/L	100	1130 mg/L
TeCC solution	260	50 ppm	60	50 ppm

Results and Discussion

Partial Chemical Oxidation of PCP. Sequential Fenton's Reagent addition to 1 Mm PCP solutions caused stepwise reductions in PCP and gradual chloride ion production. These results are presented in Figures 1 and 2, respectively. Fenton's Reagent reactivity in the dosing sequence is shown in Figure 3. These results indicate that the

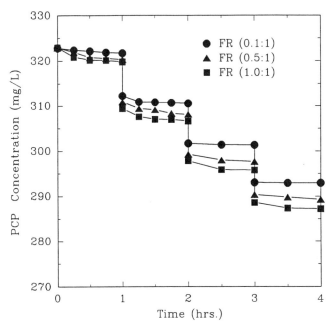

FIGURE 1. PCP Concentration Profiles during Fenton's Reagent Sequential Dosing at 20 °C

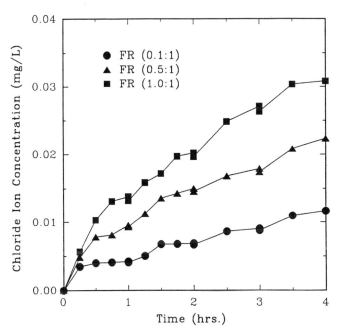

FIGURE 2. Production of Chloride Ions due to Fenton's Reagent Doses at 20 °C and pH=3.5.

chemical oxidation pretreatment with the Fenton's Reagent doses tested in this study did not induce the complete chemical oxidation of PCP. Instead, at the low concentration doses tested, calculations revealed that approximately one chloride ion per PCP molecule was produced. This result provided evidence that hydroxyl radical was produced as shown in Equation 2, and reacted subsequently with PCP as proposed in Equations 11 and 12.

$$PCP \quad + \quad \cdot OH \quad \text{-------} > \quad \text{(structures)} \quad \text{or} \quad \text{(structure)} \quad (11)$$

$$\text{(structure)} \quad \text{or} \quad \text{(structure)} \quad + \quad \cdot OH \quad \text{------} >$$

$$\text{(structure)} \quad \text{or} \quad \text{(structure)} \quad + \quad Cl^- \quad (12)$$

(THQ)   (TeCC)

Confirmation of these observations was carried out by solvent extraction, acetylation and GC/MS analysis. Two peaks, A and B were identified as acetylated components, one the residual PCP, and the other, tetrachlorocatechol. This result agreed with the ortho mechanism of PCP chemical oxidation shown in Equations 11 and 12. The production of TeCC dominated the Fenton's Reagent reactions, and alternative side reactions shown in Equations 13 through 16 could not be detected.

$$Fe^{3+} + H_2O_2 \longrightarrow Fe^{2+} + HO_2\cdot + H^+ \qquad (13)$$

$$Fe^{2+} + \cdot OH \longrightarrow Fe^{3+} + OH^- \qquad (14)$$

$$Fe^{3+} + HO_2\cdot \longrightarrow Fe^{2+} + O_2 + H^+ \qquad (15)$$

$$\cdot OH + H_2O_2 \longrightarrow H_2O + HO_2\cdot \qquad (16)$$

Overall results of the partial chemical oxidation experiments using Fenton's Reagent demonstrated that sequential chain reactions shown in Equations 2, 11, and 12 were successfully carried out to the extent intended, and the undesirable parallel reaction shown in Equations 13 -16, which dissipate the generated hydroxyl radical, had been minimized.

Biodegradation of Untreated PCP CONTROL. Sequential PCP removals by the selected microbial consortium are presented in Table 3. Also presented there are the specific PCP removal rates calculated from Equation 4 and the corresponding measured chloride ion concentrations. These results indicate that acclimated PCP-degrading SMC removed PCP from an initial 1 Mm concentration very slowly after approximately a 4-day time lag. Biological production of chloride ions demonstrated that the PCP disappearance was due to microbial degradation, not to biosorption. The 12 day aerobic biodegradation by PCP-degrading SMC resulted in a concentration decrease of only 6.4 ppm (0.024 Mm). Measured cumulative oxygen uptake rates and microbial growth rates confirmed the low level of PCP biodegradation in the CONTROL system.

Table 3. Biological PCP removals, specific PCP uptake rates and production of chloride ions by PCP-degrading SMC in the CONTROL system

Time (day)	Cumulative PCP Decrease (Mm)	Rate (day^{-1})	Cumulative Cl$^-$ Increase (Mm)
2	0.006	0.0067	0.006
4	0.010	0.0059	0.010
6	0.019	0.0071	0.023
8	0.021	0.0058	0.031
10	0.023	0.0051	0.042
12	0.024	0.0044	0.048

PCP Biodegradation Following Fenton's Reagent Partial Oxidation. PCP removals and corresponding chloride ion production are presented in Table 4 for the CONTROL and the FRSMC systems at all Fenton's Reagent doses tested. Increases in chloride ion concentrations demonstrated that the disappearance of PCP was due to microbial degradation, not due to biosorption. The microbial production of chloride ions in systems pretreated by Fenton's Reagent did not correspond in time frame to the decrease in PCP concentration.

The lack of correspondence was due to the time required to release chloride ions by microbial dehalogenases (*49*). Knackmuss (*50*) also observed that bacteria degrading chlorophenols first opened the dihydroxylated aromatic ring before producing chlorides. Chaudhry and Huang (*51*) observed one chloride ion release prior to intradiol ring cleavage; from the resulting chlorinated aliphatic fragment, another chloride ion was later removed. Pignatello *et al.* (*25*) and Lynch and Hobbie (*30*) found similar results supporting the poor correspondence of the decrease in aqueous PCP concentration and the production of chloride ions within the same time frame. In this research, release of chlorides due to microbial PCP catabolism agreed with the above observations of others cited. Production of chloride ions from PCP biodegradation was significantly increased after 6 days of microbial activity; therefore, total removal of chloride can not be expected immediately at the onset of chlorophenol biodegradation.

Table 4. Decrease in PCP concentration and production of chloride ion in FRSMC and systems

Time (day)	CONTROL PCP (Mm)	CONTROL Cl⁻ (Mm)	FRSMC(0.1:1) PCP (Mm)	FRSMC(0.1:1) Cl⁻ (Mm)	FRSMC(0.5:1) PCP (mM)	FRSMC(0.5:1) Cl⁻ (mM)	FRSMC(1.0:1) PCP (mM)	FRSMC(1.0:1) Cl⁻ (mM)
0	0.0	0.0	0.0	0.0	0.0	0.0	0.0	0.0
2	0.006	0.006	0.018	0.020	0.021	0.030	0.024	0.047
4	0.010	0.010	0.036	0.020	0.037	0.044	0.044	0.066
6	0.019	0.023	0.041	0.031	0.043	0.074	0.051	0.011
8	0.021	0.031	0.046	0.039	0.048	0.100	0.055	0.012
10	0.023	0.042	0.048	0.060	0.054	0.105	0.058	0.140
12	0.024	0.048	0.050	0.060	0.056	0.120	0.062	0.141

The specific substrate uptake rates of aqueous PCP by PCP-degrading SMC in the systems pretreated by Fenton's Reagent are presented in Table 5 and in Figure 4. The highest specific PCP uptake rate by PCP-degrading SMC following partial chemical oxidation by Fenton's Reagent, 0.164 day⁻¹, was 23 times greater than that of the CONTROL system.

Table 5. Specific PCP uptake rates (day⁻¹) by PCP-degrading SMC for systems pretreated by Fenton's Reagent and for the CONTROL

Time (day)	CONTROL	FRSMC (0.1:1)	FRSMC (0.5:1)	FRSMC (1.0:1)
0	0.0	0.0	0.0	0.0
2	0.0067	0.137	0.155	0.164
4	0.0059	0.092	0.098	0.105
6	0.0071	0.059	0.064	0.068
8	0.0058	0.045	0.048	0.052
10	0.0051	0.036	0.039	0.042
12	0.0044	0.030	0.033	0.035

The maximum specific rate of PCP uptake appeared at the beginning of the biodegradation period for systems pretreated by Fenton's Reagent, as presented in Figure 4. Miller and co-workers (52) measured the enhancement of biodegradation of dichlorophenol (DCP) and trichlorophenol (TCP) by photolyzed hydrogen peroxide and reported similar significant biological degradation during the first two days. In addition, this study demonstrated that the specific PCP uptake rate was proportional

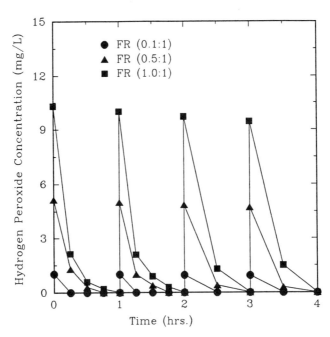

FIGURE 3. Hydrogen Peroxide Profiles during Fenton's Reagent Sequential Dosing at 20 °C.

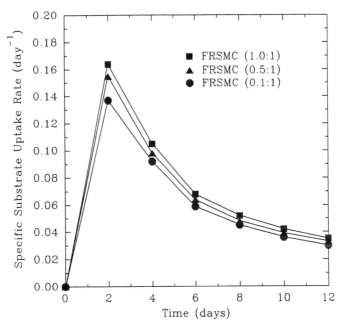

FIGURE 4. FRSMC Specific PCP Uptake Rates during 12-day Biodegradation Period.

to the molar ratio of hydrogen peroxide in Fenton's Reagent to PCP concentration, 0.137 day^{-1} for 0.1:1, 0.155 day^{-1} for 0.5:1 and 0.164 day^{-1} for 1:1. These results of FRSMC systems were subjected to a t-test and were significantly different at a 0.05 probability level (t_{calc}=2.45 > t_{table}=1.782 for comparison of 0.1:1 with 0.5:1, t_{calc}=2.93 > t_{table}=1.782 for comparison of 0.1:1 with 1.0:1). The highest specific PCP uptake rate with PCP-degrading SMC for the non-pretreated system was 0.0071 day^{-1} between days 4 and 6. Following day 6, the rate decreased to a rate approximately 60% of maximum. The maximum specific PCP uptake rates in the systems pretreated by Fenton's Reagent were approximately twenty times those in the non-pretreated system and were measured between time 0 and day 2. In addition, specific PCP uptake rates for the systems pretreated by Fenton's Reagent were still approximately one order of magnitude higher at the end of 12 days than the comparable values of the non-pretreated system.

The explanations for enhancement of the biodegradation rate are as follows: 1.) that the PCP partial oxidation products were less toxic substrates to the microorganisms than was the parent PCP, and these PCP partial oxidation products induced enhanced degradation by co-metabolic mechanisms, or 2.) that the environment provided was less toxic to the microorganisms due to lowered initial PCP concentration. The enhanced biodegradation following Fenton's Reagent pretreatment, was not due to the slight reduction in initial PCP concentration from partial chemical oxidation pretreatment, only 2.9%. The mechanism for the enhancement of PCP biodegradation, therefore, must have been due to co-metabolism, as described above. This probability was substantiated by the observation that after four days of enhanced PCP uptake, the oxidation products which enhanced the co-metabolic uptake were probably decreased to a low concentration, and biodegradation rates decreased from a level twenty times greater than the comparable CONTROL system value to a rate approximately half the maximum value. The rates of PCP uptake were corroborated by rates of chloride ion production, microbial growth and cumulative oxygen uptake.

The growth of PCP-degrading SMC in the systems pretreated by Fenton's Reagent are presented in Figure 5. The microbial growth by PCP-degrading SMC following chemical oxidation by Fenton's Reagent was higher than that in the CONTROL system and was proportional to the Fenton's Reagent doses. Furthermore, the periods of highest growth of PCP-degrading SMC corresponded to the period of highest PCP biodegradation.

The gradual production of chloride ions for the PCP system pretreated by Fenton's Reagent is shown in Figure 6. The production of chloride ion also was proportional to the Fenton's Reagent dose. The cumulative oxygen uptake by PCP-degrading SMC in Fenton's Reagent pretreated system is presented in Figure 7. The cumulative oxygen uptake by PCP-degrading SMC observed in the system pretreated by Fenton's Reagent was proportional to the Fenton's Reagent dose. All results show that pretreatment of PCP had an important enhancing effect on the activity of PCP-degrading SMC.

Mechanism of Biodegradation Enhancement. The microbial growth response to the toxic chemicals was related to the concentration of toxic chemicals. Moreover, the individual utilization of toxic chemicals for energy or growth can be determined from the comparison of specific substrate uptake rate and growth rate. Specific PCP uptake rates, K, were determined in batch experiments with varying initial PCP concentrations and fitted to the Michaelis-Menten model as described.

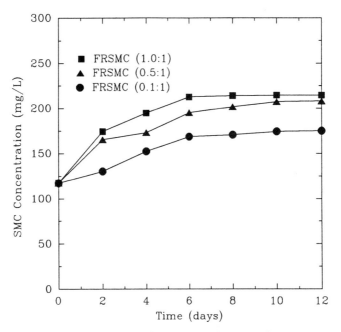

FIGURE 5. Micobial Growth of PCP-degrading SMC in FRSMC System.

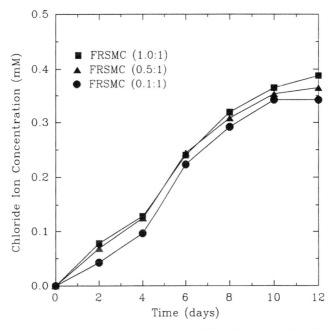

FIGURE 6. PCP-degrading SMC Production of Chloride Ions During 12-day Period.

Results are presented graphically in Figure 8 and numerically in Table 6. The values of k and K_m for the non-pretreated aqueous PCP system were found to be 0.023 day^{-1} and 180 mg/L, respectively. And, the corresponding values for the pretreated

Table 6. Specific PCP uptake rate constants and biokinetic constants for the CONTROL and the system pretreated by Fenton's Reagent

Non-pretreated system		Pretreated system	
S_o (mg/L)	K (day^{-1})	S_o (mg/L)	K (day^{-1})
5.9	9.6×10^{-4}	3.6	9.0×10^{-3}
17.0	9.6×10^{-4}	11.2	1.0×10^{-2}
26.4	2.4×10^{-3}	16.6	1.2×10^{-2}
50.4	2.1×10^{-3}	40.7	1.9×10^{-2}
104.2	1.9×10^{-3}	65.2	3.1×10^{-2}
141.2	1.2×10^{-2}	86.0	6.1×10^{-2}
154.7	2.2×10^{-2}	102.0	5.3×10^{-2}
243.3	6.7×10^{-3}	146.0	6.0×10^{-2}
		190.0	5.0×10^{-2}

Non-pretreated system		Pretreated system	
k (day-1)	K_m (mg/L)	k (day^{-1})	K_m (mg/L)
*2.3×10^{-2}	180	*8.5×10^{-2}	79

* confidence level $< \pm 5\%$

system were 0.085 day^{-1} and 79 mg/L, respectively. The confidence level associated with each maximum substrate uptake rate constant in Table 7 is less than 5%, significantly less than the acceptable error associated with standard BOD determinations. Furthermore, Michaelis-Menten constants were subjected to a t-test to determine whether the resulting constants for the non-pretreated and pretreated systems were the same or significantly different at a probability level equal to 0.05. The values of k and K_m from the non-pretreated system, and values of k and K_m from the pretreated system were significantly different (t_{calc}=2.02 > t_{table}=1.671 at α=0.05).

Michaelis-Menten constants from the non-pretreated and pretreated system showed that PCP uptake by PCP-degrading SMC was enhanced when the PCP aqueous system was enhanced by Fenton's Reagent partial oxidation. The maximum substrate uptake rate for the pretreated system was increased four times greater than that of the unpretreated system, from 0.023 day^{-1} to 0.085 day^{-1}. Furthermore, K_m was reduced by approximately one half when the PCP system was pretreated by Fenton's Reagent, from 180 mg/L for non-pretreated system to 79 mg/L for pretreated system.

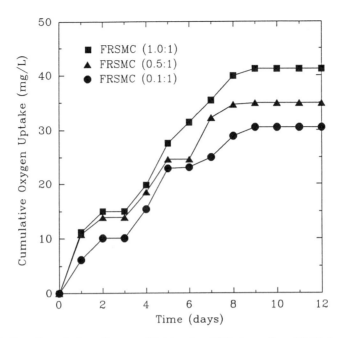

FIGURE 7. Cumulative Oxygen Uptake by PCP-degrading SMC in FRSMC Systems.

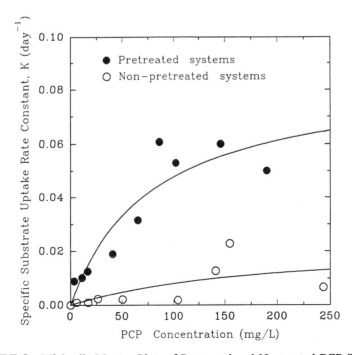

FIGURE 8. Michaelis-Menten Plots of Pretreated and Nontreated PCP Systems.

The values for the Haldane biokinetic growth rate constants μ_{max}, K_s and K_i were determined in batch experiments by obtaining the specific growth rates, μ, at several different initial PCP concentrations and calculated using Equation 6. These data were fitted to Equation 8 using the software described and subsequent biokinetic constants, S^*, the critical substrate concentration and μ^*, the peak growth rate, were calculated from Equations 9 and 10, respectively.

In this study the growth rate for the non-pretreated system and for the pretreated system increased with increasing PCP concentration up to the critical substrate concentration, S^*. These results are presented in Figure 9 and illustrate the critical substrate concentration for the non-pretreated and pretreated systems of 22.2 mg/L and 29 mg/L, respectively. The growth rate decreased significantly above this concentration in both the non-pretreated and the pretreated systems. Not shown in Figure 9 is that the time to reach the maximum specific growth rate for the system pretreated by Fenton's Reagent was 1 to 2 days; for the non-pretreated system, 2 to 4 days were required to reach the maximum specific growth rates. This result was corroborated by the increase in K_i due to Fenton's Reagent pretreatment, almost 5 times greater than the CONTROL system. Calculated biokinetic constants for non-pretreated and pretreated PCP systems are presented in Table 7.

Table 7. Biokinetic constants for PCP-degrading SMC in CONTROL and in systems pretreated by Fenton's Reagent

System	μ_{max} (day^{-1})	K_s (mg/L)	K_i (mg/L)	μ^* (day^{-1})	S^* (mg/L)
Non-pretreated	0.17	26.9	18.4	[a]0.050	22.2
Pretreated	0.19	9.7	87.0	[a]0.113	29.0

[a] : confidence level $< \pm 5\%$

For the pretreated system, the values of μ_{max}, K_s, and K_i, were 0.19 day^{-1}, 9.7 mg/L and 87.0 mg/L, respectively. For the non-pretreated system, μ_{max}, K_s and K_i values were 0.17 day^{-1}, 26.9 mg/L and 18.4 mg/L, respectively. All these values indicate that the acclimated PCP-degrading SMC benefitted from chemical oxidation pretreatment. All of these results indicate that the enhanced growth rate was due to the generation of co-substrates by Fenton's Reagent partial oxidation, and the decrease in toxicity was due to the presence of these co-substrates. The decrease in growth rate at PCP concentrations greater than S* demonstrated that high concentrations of PCP were increasingly toxic to the acclimated PCP-degrading SMC. The fact that the growth rate at high PCP concentrations was greater than in the untreated system was due not to the slight decrease in PCP concentration at these Fenton's Reagent doses, but due to the presence of the oxidation products. At high PCP concentrations, the uptake was maintained at a high rate, increasing with increasing PCP concentration. Since the Fenton's Reagent dose was administered at a constant molar ratio, at high

PCP concentrations, the Fenton's Reagent dose was larger, and the resulting concentration of generated oxidation products was larger. The large concentration of PCP oxidation products present enhanced the PCP uptake, but the high concentration of PCP inside the acclimated cells inhibited the microbial growth at PCP concentrations greater than S^*. This result demonstrated that the oxidation products served as co-substrates which enhanced PCP uptake and microbial growth at low PCP concentrations, but at high PCP concentrations, the presence of high concentrations of PCP oxidation products increased the PCP uptake to an increasingly inhibitory level inside the cell. The PCP oxidation products must, therefore, enhance the membrane transport of PCP; the higher the concentration of both PCP and its oxidation products, the larger the uptake rate due to enhancement of membrane transport.

For the pretreated systems, the values of μ^* and S^* were 0.113 day^{-1} and 29 mg/L, respectively, whereas μ^* and S^* values for the non-pretreated system were 0.050 day^{-1} and 22.2 mg/L, respectively. The critical substrate concentration, S^*, and the peak growth rate, μ^* were enhanced in the pretreated system. The value of μ^* was twice as great for the pretreated system than that for the non-pretreated system. The confidence level associated with each peak growth rate is presented in Table 7 and is less than $\pm$ 5 %, significantly less than the acceptable error associated with standard BOD determinations.   Several researchers reported that mixtures of organic compounds influenced the biokinetics of microbial growth compared with single organic substrate as the sole carbon source (*53,-55, 35*). Lynch and Hobbie (*30*) claimed that co-substrates are necessary to support the growth of the co-metabolizing microorganisms since co-metabolic transformations of xenobiotic compounds are not directly linked to their utilization of energy sources. Therefore, the co-substrate may take part directly in co-metabolism; the co-substrate may be responsible for the expression of the co-metabolic pathway, or it may supply co-factors for the enzymes of the pathway or energy for the transport of the xenobiotics.   This addition of co-substrates to enhance the growth of PCP-degrading microorganisms implies that the treatability of PCP with microorganisms can be improved by partial oxidation pretreatment to produce such co-metabolites. To examine this possibility further, PCP uptake experiments were conducted to examine the co-metabolic mechanisms. PCP solutions were spiked with TeCC or with solutions resulting from the oxidation pretreatment with Fenton's Reagent, as defined in Table 2. In these experiments, SYSTEM 1 received 15 ppm TeCC in 75 ppm PCP solution, and SYSTEM 2 received 5 ppm TeCC in 100 ppm PCP solution for the initial condition of PHASE I. PHASE II represents the period following a 150 ml addition of PCP solution pretreated by Fenton's Reagent, and PHASE III represents the period following another 100 ml addition of PCP solution pretreated by Fenton's Reagent. Results of these sequential experiments are shown in Figure 10 and biodegradation rate constants are presented in Table 8.

Biodegradation rate constants from each PHASE in each system were analyzed by a t-test to determine how biodegradation rate constants in each PHASE of SYSTEM 1 and SYSTEM 2 compared to the CONTROL system. These results are presented in Table 9. For SYSTEM 1, receiving 15 ppm of TeCC, the biodegradation rate constant of PHASE I was Table 8. Specific PCP uptake rate constants (day^{-1}) of systems supplemented with TeCC and subsequently with partially oxidized PCP solutions and comparable constants of CONTROL system.

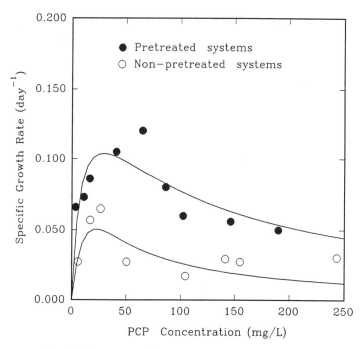

FIGURE 9. Haldane Plots of Pretreated and Nontreated PCP Systems.

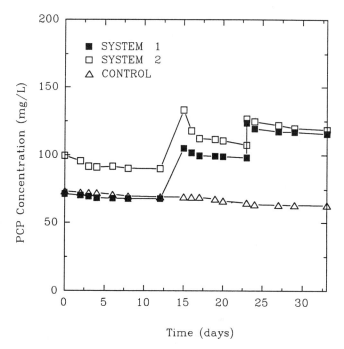

FIGURE 10. PCP Concentration Profiles during Mechanistic Spiking Experiments.

Table 8. Specific PCP uptake rate Constants (day^{-1}) for TeCC Supplemented and Partially Oxidized Systems

	SYSTEM 1	SYSTEM 2	CONTROL
PHASE I (overall)	0.023	0.035	0.030
DAY 0 - DAY 4	0.049	0.084	0.036
DAY 6 - DAY 12	0.028	0.026	0.021
PHASE II (overall)	0.045	0.189	0.042
DAY 15 - DAY 17	0.068	0.394	0.039
DAY 19 - DAY 23	0.023	0.079	0.043
PHASE III (overall)	0.038	0.072	0.008
DAY 23 - DAY 26	0.093	0.102	0.020
DAY 27 - DAY 33	0.011	0.046	0.002

SYSTEM 1: The system receiving 15 ppm TeCC in 75 ppm PCP solution
SYSTEM 2: The system receiving 5 ppm TeCC in 100 ppm PCP solution

For SYSTEM 1 which had an addition of 15 ppm TeCC, the biodegradation rate constant was 0.023 day^{-1}; this rate constant was less than that for the CONTROL, indicating that the high concentration of TeCC was inhibitory. When the PCP solution pretreated by Fenton's Reagent was spiked into the same system during phase II, the biodegradation rate was enhanced to 0.045 day^{-1}, almost twice the value from PHASE I.

Table 9. The results of t-test for the biodegradation rate constants compared with CONTROL system

	SYSTEM 1		SYSTEM 2	
	t_{calc}	t_{table}	t_{calc}	t_{table}
PHASE I	2.28**	1.782	1.22*	1.782
PHASE II	1.41*	1.782	2.54**	1.782
PHASE III	3.83**	1.782	3.77**	1.782

* : Biodegradation rate is not significantly different from the CONTROL.
** : Biodegradation rate is significantly different from the CONTROL.

Supplemental addition of 5 ppm tetrachlorocatechol in PHASE I of SYSTEM 2 exhibited slight PCP biodegradation enhancement compared to the CONTROL system, and the biodegradation rate of PHASE II in SYSTEM 2 was five times higher than that in PHASE I. These results indicate that TeCC was not the major rate-enhancing PCP oxidation product. The lower biodegradation rate in PHASE III can be explained by 1.) elevated PCP concentration, and 2.) diluted biomass concentration due to spiking with 100 ml of oxidized PCP solution.

The PCP-degrading SMC community used in this study was comprised of *Pseudomonas putida* and *Pseudomonas aeruginosa*. Even the acclimated PCP-degrading SMC needed co-metabolic oxidation products to enhance the transport of PCP through the cell membrane. In the pretreated FRSMC systems this enhanced uptake was maintained at high initial concentrations when corresponding microbial growth was inhibited at PCP concentrations larger than S*.

Conclusions

1. Partial chemical oxidation of 1 mM PCP solutions by the highest dose of Fenton's Reagent resulted in a 2.9% reduction of PCP concentration. Hydroxyl radical, the strong oxidizing agent generated from hydrogen peroxide in the presence of ferrous ion catalyst, was efficiently generated and utilized to partially oxidize PCP. The hydrogen peroxide present in Fenton's Reagent was very reactive and disappeared within the four-hour chemical oxidation period. On a molar basis, the efficiency of PCP dechlorination and consequent chloride ion production by Fenton's Reagent was approximately equal to the decrease in PCP concentration. The partial oxidation of PCP pretreated by Fenton's Reagent doses used in this study resulted in oxidation products with fewer chlorine substituents than the parent model chemical, and their production in the presence of the residual parent chemical provided appropriate conditions for observing subsequent enhancement of microbial biodegradation.

2. Biodegradation of PCP by acclimated SMC was slow in the CONTROL system. The 1 mM concentrations of model chemicals used in this study were inhibitory to SMC. The biodegradation rates of PCP in the system pretreated by Fenton's Reagent were more rapid and more extensive than those in the CONTROL system. Fenton's Reagent partial oxidation pretreatment also diminished time lags observed for SMC subsequent biodegradation. The enhanced removal rate of PCP following Fenton's Reagent partial oxidation was only temporary and continued until apparently all the co-metabolized oxidation products were removed. Then the removal rate reverted back to a rate approximately twice the level of the non-pretreated CONTROL system. The enhanced biodegradation rates of PCP following Fenton's Reagent partial oxidation pretreatment were correlated with enhanced microbial chloride ion production due to enzymatic dechlorination and with enhanced microbial growth and cumulative oxygen uptake. The decrease in K_s due to Fenton's Reagent pretreatment demonstrated a reduction in toxicity provided for acclimated SMC by reducing the 1 mM initial PCP concentration by only 2.9%. This reduction in toxicity must be due to the presence of oxidation products and their apparent enhancement of membrane transport. In addition, the specific growth rate of PCP degrading SMC was enhanced by pretreatment with Fenton's Reagent, compared to the inhibitory effect exerted on the SMC in the non-pretreated CONTROL system. In addition, the maximum growth rate, μ_{max}, and μ^* were increased for the pretreated systems, and the critical substrate concentration, S^*, was reduced when the systems were pretreated. In addition, at concentrations of PCP higher than S^*, pretreated PCP supported more SMC growth than in the non-pretreated system. Selected microbial consortium increases in microbial growth rates due to chemical oxidation pretreatment of parent model chemicals was due to both reduction in toxicity of PCP in the presence of its partial chemical oxidation products, since Haldane inhibitory constants were reduced, and due

to co-metabolism, since the substrate uptake rates were enhanced at all substrate concentrations examined.

3. In the case of PCP systems where supplemental chlorinated organic chemical was added, the biodegradation rates in systems where a high concentration of TeCC was added inhibited biodegradation; whereas the system with a low concentration of TeCC supplement demonstrated a slightly enhanced biodegradation rate. Subsequent spiking with PCP solution pretreated by Fenton's Reagent enhanced the biodegradation rate more than in the system in which specific supplemental chemical was added. The tetrachlorocatechol was not the principal enhancing co-substrate produced by partial PCP chemical oxidation.

4. Overall, pretreatment by Fenton's Reagent enhanced the biodegradation rates of PCP and the microbial growth rate. The mechanism for the enhancement of PCP biodegradation was due to a reduction in toxicity of PCP in the presence of the partial chemical oxidation products and due to co-metabolism of the PCP and these oxidation products.

Acknowledgments

This work was funded by EPA Grant No. NPIR 002-88. The work of ICI Pharmaceuticals is gratefully acknowledged for their bacterial identification expertise.

Literature Cited

1. United States Environmental Protection Agency. Guidelines Establishing Test Procedures for the Analysis of Pollutants. *Federal Register.* **1979**, *44*, pp. 233.

2. Crosby, D. G. Environmental Chemistry of Pentachlorophenol. *Pure Appl. Technol.* **1984**, *53*, pp. 1051-1080.

3. Godsy, E. W., Georlitz, D. F. and Grabic-Galic, D. Anaerobic Biodegradation of Creosote Contaminants in Natural and Simulated Ground Water Ecosystem, In: *EPA Symposium on Bioremediation of Hazardous Wastes; EPA's Biosystems Technologies Development Program,* Arlington, VA, 1990; pp. 31-33.

4. Goerlitz, D. F., Trouman, D. E., Godsy, E. M. and Franks, B. J. Migration of Wood-Preserving Chemicals in Contaminated Groundwater in a Sand Aquifer at Pensacola, Florida. *Environ. Sci. Technol.* **1985**, *19*, pp. 955-961.

5. Thompson, D. L. K. and Kaiser, K. L. E. Quantitative Structure-Toxicity Relationship of Halogenated Phenols on Bacteria. *Bull. Environm. Contam. Toxicol.* **1982**, *29*, pp. 130-136.

6. Brummett, T. B. and Ordal, G. W. Inhibition of Amino Acid Transport in *Bacillus subtilis* by Uncouplers of Oxidative Phosphorylation. *Arch. Biochem. Biophys.* **1977**, *178*, pp. 368-372.

7. Saber, D. L. and Crawford, R. L. Isolation and Characterization of *Flavobacterium* Strains that Degrade Pentachlorophenol. *Appl. Environ. Microbiol.* **1985**, *50*, pp. 1512-1518.

8. Shore, R. F., Myhill, D. G., French, M. C., Leach, D. V. and Stebbings, R. E. Toxicity and Tissue Distribution of Pentachlorophenol and Permethrin in Pipistrelle Bats Experimentally Exposed to Treated Timber. *Environ. Pollu.* **1991**, *73*, pp. 101-118.

9. Renberg, L. E., Marell, G., Sundstorm, G. and Adolfsson-Erici, M. Levels of Chlorophenols in National Waters and Fish after an Accidental Discharge of a Wood-Impregnating Solution. *Ambio* **1983**, *12*, pp. 121-123.

10. Moos, L. P., Kirsh, E. J., Wukasch, R. F. and Grady, Jr., C. P. L. Pentachlorophenol Biodegradation I; Aerobic. *Water Res.* **1983**, *17*, pp. 1575-1584.
11. Melcer, H. and Bedford, W. K. Removal of Pentachlorophenol in Municipal Activated Sludge Systems. *J. Water Pollut. Control Fed.* **1988**, *60*, pp. 622-626.
12. Kuwatsuka, S. and Igarashi, M. Degradation of PCP in Soil. *Soil Sci. Plant Nutri.* **1975**, *21*, pp. 405-414.
13. Watanabe, I. Pentachlorophenol-Decomposing and PCP-Tolerant Bacteria in Field Soil Treated with PCP. *Soil Biol. Biochem.* **1976**, *9*, pp. 99-103.
14. Kaufman, D. Degradation of PCP in Soil, and by Soil Microorganisms. In:*Pentachlorophenol.* Rao, K. R., Ed.; Plenum Press, New York, NY, 1978.
15. Baker, M. D. and Mayfield, C. I. Microbial and Nonbiological Decomposition of Chlorophenols and Phenol in Soil. *Water Air Soil Pollut.* **1980**, *13*, pp. 411-424.
16. Edgehill, R. U. and Finn, R. K. Microbial Treatment of Soil to Remove Pentachlorophenol. *Appl. Environ. Microbiol.* **1983**, *45*, pp. 1122-1125.
17. Smith, J. A. and Novak, J. T. Biodegradation of Chlorinated Phenols in Subsurface Soils. *Water Air Soil Pollut.* **1987**, *33*, pp. 29-42.
18. Brown, E., Pignatello, J. T., Martinson, M. M. and Crawford, R. L. Pentachlorophenol Degradation : A Pure Bacterial Culture and an Ephilic Microbial Consortium. *Appl. Environ. Microbiol.* **1986**, *52*, pp. 92-97.
19. Chu, J. P. and Kirsch, E. J. Metabolism of Pentachlorophenol by an Axenic Bacterial Culture. *Appl. Environ. Microbiol.* **1972**, *23*, pp. 1033-1035.
20. Stanlake, G. J. and Finn, R. K. Isolation and Characterization of a Pentachlorophenol-Degrading Bacterium. *Appl. Environ. Microbiol.* **1982**, 44:1421-1427 (1982).
21. Salkinoja-Salonen, M. S., Hakulinen, R., Valo, R. and Apajalaht, J. Biodegradation of Recalcitrant Organochlorine Compounds in Fixed Film Reactors. *Wat. Sci. Tech.* **1983**, *15*, pp. 309-319.
22. Steiert, J. G., Pignatello, J. T. and Crawford, R. L. Degradation of Chlorinated Phenols by a Pentachlorophenol-Degrading Bacterium. *Appl. Environ. Microbiol.* **1987**, *53*, pp. 907-910.
23. Chu, J. P. and Kirsch, E. J. Utilization of Halophenols by a Pentachlorophenol Metabolizing Bacterium. *Developments in Industrial Microbiology.* **1973**, *14*, pp. 264-273.
24. Reiner, E. A. and Kirsh, E. J. Microbial Metabolism of Pentachlorophenol. In:*Pentachlorophenol.* Rao, K. R., Ed.; Plenum Press. New York, NY, 1978.
25. Pignatello, J. J., Martinson, M. M., Steiert, J. G., Carlson, R. E. and Crawford, R. L. Biodegradation and Photolysis of Pentachlorophenol in Artificial Freshwater Streams. *Appl. Environ. Microbiol.* **1983**, *46*, pp. 1024-1031.
26. Volskay, V. T. and Grady, Jr., C. P. L. Toxicity of Selected RCRA Compounds to Activated Sludge Microorganisms. *J. Water Pollut. Control Fed.* **1988**, *60*, pp. 1850-1856.
27. Crawford, R. L. and Mohn, W. W. Microbiological Removal of Pentachlorophenol from Soil Using *Flavobacterium. Enzyme Microb. Technol.* **1985**, *7*, pp. 617-620.
28. Steiert, J. G. and Crawford, R. L. Catabolism of Pentachlorophenol by *Flavobacterium* sp. *Biochem. Biophys. Res. Commun.* **1986**, *141*, pp. 825-830.
29. Rochkind-Dubinsky, M. L., Sayler, G. S. and Blackburn, J. W. Chlorophenols In: *Microbiological Decomposition of Chlorinated Aromatic Compounds.* Rochkind-

Dubinsky, M. L., Sayler, G. S. and Blackburn, J. W., Eds.; Marcel Dekker, Inc., New York, NY, 1987.

30. Lynch, J. M. and Hobbie, J. E. The Problem of Xenobiotics and Recalcitrance. In: *Micro-organisms in Action*. Lynch, J. M. and Hobbie, J. E., Eds.; Blackwell Scientific Pub., Cambridge, MA, 1988.

31. Alexander, M. Biodegradation of Chemicals of Environmental Concern. *Science* **1981**, *211*, pp. 132-138.

32. Rosenberg, A. and Alexander, M. 2,4,5-Trichlorophenoxyacetic Acid (2,4,5-T) Decomposition in Tropical Soil and Its Cometabolism by Bacteria in Vitro. *J. Agric. Food Chem.* **1980**, *28*, pp. 705-709.

33. Karns, J. S., Kilbane, J. J., Duttagupta, S. and Chakrabarty, A. M. Metabolism of Halophenols by 2,4,5-Trichlorophenoxyacetic Acid-Degrading *Pseudomonas cepacia*. *Appl. Environ. Microbiol.* **1983**, *46* pp. 1176-1181.

34. Kohler, H. P. E., Kohler-Staub, D. and Focht, D. D. Cometabolism of Polychlorinated Biphenyl: Enhanced Transformation of Aroclor 1254 by Growing Bacteria Cells. *Appl. Environ. Microbiol.* **1988**, *54*, pp. 1940-1945.

35. Klecka, G. M. and Maier, W. J. Kinetics of Microbial Growth on Mixtures of Pentachlorophenol and Chlorinated Aromatic Compounds. *Biotechnol. Bioeng.* **1988**, *31*, pp. 328-335.

36. Thompson, D. L. K. and Strachan, W. M. J. Biodegradation of Pentachlorophenol in a Simulated Aquatic Environment. *Bull. Environm. Contam. Toxicol.* **1981**, *26*, pp. 85-90.

37. Hendriksen, H. V., Larsen, S. and Anring, B. K. Influence of a Supplemental Carbon Source on Anaerobic Dechlorination of Pentachlorophenol in Granular Sludge. *Appl. Environ. Microbiol.* **1992**, *58*, pp. 365-370.

38. Fenton, H. J. H. Correspondence on a New Reaction of Tartaric Acid. *Chemical News J. Physical Science*. **1876**, *33*, pp. 190.

39. Haber, F. and Weiss, J. The Catalytic Decomposition of Hydrogen Peroxide by Iron Salts. *Proc. R. Soc. A*, **1934**, *147*, pp. 332-351.

40. Cohen, G. The Fenton Reaction. In: *Handbook of Methods for Oxygen Radical Research*. Greenwald, R. A., Ed.; CRC Press, Boca Raton, FL, 1989.

41. Schumb, W. C., Satterfield, C. N. and Wentworth, R. L. *Hydrogen Peroxide*. Reinhold Pub., Co. New York, NY, 1955.

42. Barbeni, M., Minero, C. and Pelizzetti, E. Chemical Degradation of Chlorophenols with Fenton's Reagent. *Chemosphere*. **1987**, *16*, pp. 2225-2237.

43. Watts, R. J., Leung, S. W. and Udell, M. D. Treatment of Contaminated soils Using Catalyzed Hydrogen Peroxide Presented at *Chemical Oxidation Technology for the Nineties First International Symposium*, Vanderbilt University, Nashville, TN, 1991.

44. Bishop, D. F., Stern, G., Fleischman, M. and Marshall, L. S. Hydrogen Peroxide Catalytic Oxidation of Refractory Organics in Municipal Wastewaters. *Ind. Eng. Chem.: Process Design and Development*. **1968**, *7*, pp. 110-117.

45. Bowers, A. R., Gaddipati, P., Eckenfelder, Jr., W. W. and Monsen, R. M. Treatment of Toxic or Refractory Wastewaters with Hydrogen Peroxide. *Wat. Sci. Tech.* **1989**, *14*, pp. 431-486.

46. Lee, S. H. and Carberry, J. B. Biodegradation of Pentachlorophenol Enhanced by Chemical Oxidation Pretreatment. *Water Environ. Res.* **1992**, *64*, pp. 682-690.

47. Kolthoff, I. M. and Sandell, E. B. *Textbook of Quantitative Inorganic Analysis*. I.Kolthoff, M. and Sandel, E. B., Eds.; Macmillan, New York, NY, 1936.

48. Haggblom, M. M., Apajalahti, J. H. A. and Salkinoja-Salonen, M. S. Hydroxylation and Dechlorination of Chlorinated Guaiacols and Syringols by *Rhodococcus chlorophenolicus. Appl. Environ. Microbiol.* **1988**, *54*, pp 683-687.
49. Weightman, A. J., Weightman, A. L. and Slater, J. H. Stereospecificity of 2-Monochloro-propionate Dehalogenation by the Two Dehalogenases of *Pseudomonas putida* pp3: Evidence for Two Different Dehalogenation Mechanisms. *J. Gen. Microbiol.* **1982**, *128*, pp. 1755-1762.
50. Knackmuss, H. Degradation of Halogenated and Sulfonated Hydrocarbons. In: *Microbial Degradation of Xenobiotics and Recalcitrant Compounds.* Leisinger, T., Hutter, R., Cook, A. M. and Nuesch, J., Eds.; Academic Press, New York, NY, 1981.
51. Chaudhry, G. R. and Huang, G. H. Isolation and Characterization of a New Plasmid from *Flavobacterium* sp. Which Carries the Genes for Degradation of 2,4-Dichlorophenoxyacetate. *J. Bacteriol.* **1988**, *170*, pp. 3897-3902.
52. Miller, R. M., Singer, G. M., Rosen, J. D. and Bartha, E. Sequential Degradation of Chlorophenols by Photolytic and Microbial Treatment. *Environ. Sci. Technol.* **1988**, *22*, pp. 1215-1219.
53. Yoon, H., Klinzing, G. and Blanch, H. W. Competition for Mixed Substrates by Microbial Populations. *Biotechnol. Bioeng.* **1977**, *19*, pp. 1193-1210.
54. Law, A. T. and Button, D. K. Multiple-Carbon-Source-Limited Growth Kinetics of a Marine *Coryneform* Bacterium. *J. Bacteriol.* **1977**, *129*, pp. 115-123.
55. Papanastasiou, A. C. and Maier, W. J. Kinetics of Biodegradation of 2,4-Dichlorophenoxyacetate in the Presence of Glucose. *Biotechnol. Bioeng.* **1982**, *24*, pp. 2001-2011.

RECEIVED January 14, 1994

Chapter 14

Oxidation of s-Triazine Pesticides

Cathleen J. Hapeman

Environmental Chemistry Laboratory, Agricultural Research Service, U.S. Department of Agriculture, Beltsville, MD 20705

A study of the aqueous ozonation of 2-chloro-s-triazine pesticides, atrazine, simazine and propazine, is presented. The ozonation process gave rise to a mixture of products whose structure and abundance is dependent on the duration of ozonation. Structures were determined by chemical methods, HPLC, mass and NMR spectroscopy. In all the primary and secondary ozonation products oxidation of the N-alkyl to the N-acetyl and/or removal of the amino substituents occurred and the s-triazine ring remained intact. Dechlorination was not observed. Reaction of the amide was not found if an alkyl functionality was present. A proposed degradation pathway of the s-triazines is described as well as a reaction product profile of atrazine. The relative ozonation rate of N-ethyl was found to be five times greater than the rate of N-isopropyl as determined from competitive reactions and the rates of atrazine product formation.

Pesticide rinsate from application equipment typically consists of a variety of formulating agents, surfactants, emulsifiers, fertilizers and pesticides at concentrations less than 200 ppm (1). In some regions, this material can be collected and used in subsequent applications. Environmentally safe disposal of unusable equipment rinsate and other pesticide wastes is of great concern to pesticide applicators and farmers because improper disposal has been directly linked to point source contamination of groundwater and farm wells (2,3). Furthermore, effective mineralization of pesticide residues is needed in the purification of drinking water at point-of-use in agriculture areas.

Field work has demonstrated the potential of a binary remediation process consisting of oxidation and microbial mineralization, i.e., breakdown to CO_2, H_2O, NH_3 or NO_3^-, and inorganic salts. Laboratory and field trials have shown the s-triazines, such as atrazine which is the most widely used s-triazine, to be somewhat more recalcitrant than other common herbicides, necessitating further study of both the chemical and biological degradation pathways (1). Preliminary tests indicated that 2-chloro-4,6-diamino-s-triazine (CAAT) was the principle product of atrazine ozonation (4). Furthermore, a gram positive rod, DRS-I, was found to mineralize CAAT to CO_2 and to utilize the organic nitrogen of CAAT under simulated field conditions (5).

Most ozonation investigations of dilute solutions of organic compounds in water have typically examined loss of parent material or total organic carbon (6 - 9) although a few studies have identified products (10, 11). Furthermore, ozone is not the only oxidizing species present under most aqueous ozonation conditions (12 - 15). Understanding the overall process and determining the active oxidizing species and their respective roles in the degradation of the organic compound requires that the fate of the organic species be known.

Thus, optimization of the first stage of this combined chemical-biological process requires that the ozonation products be identified, the ozonation mechanism elucidated and the limiting parameters defined. In this study, comparison of the reaction rates of atrazine and other related s-triazines, propazine (2-chloro-4,6-bis-isopropylamino-s-triazine) and simazine (2-chloro-4,6-bis-ethylamino-s-triazine), was also performed to provide details concerning the relative rates of reaction.

Materials and Methods

Chemicals. For convenience the nomenclature system used by Cook (16) is used here: A = amino, C = chloro, E = ethylamino, I = isopropylamino, O = hydroxy and T = triazine ring. Several ozonation products contained an acetamido group and in keeping with this nomenclature, D has been added to denote this moiety (Table I). Atrazine (2-chloro-4-ethylamino-6-isopropylamino-s-triazine), simazine (2-chloro-4,6-diethylamino-s-triazine), propazine (2-chloro-4,6-diisopropylamino-s-triazine), 6-amino-2-chloro-4-isopropylamino-s-triazine, 6-amino-2-chloro-4-ethylamino-s-triazine, and 2-chloro-4,6-diamino-s-triazine were gifts from CIBA-GEIGY. Ultra pure water (Modulab, Type I HPLC, Continental Water System Corp., San Antonio, TX 78229) of neutral pH was used to prepare solutions of s-triazines.

Ozonation Procedure. Ozonation experiments were carried out in one of two previously described reactors: (1) Studies involving the isolation of intermediates and elucidation of the s-triazine degradation pathway were conducted using a custom designed glass reactor where ozone/oxygen gas was passed over the top of the solution to decrease the ozone contact time with the solution decreasing the rate of reaction (17). (2) A standard photoreactor retrofitted with a bottom feed sintered glass frit was used for all kinetic experiments (18). Ozone was generated using a PCI Ozone Generator

Model GL-1B (PCI Ozone Corporation, West Caldwell, NJ 07006) with oxygen feed. Oxygen/ozone was delivered to the reactor at 1 L/min with an ozone concentration between 0.2 and 0.4 % as determined using a PCI Ozone Monitor Model HC. In pathway elucidation and isolation studies, typical initial concentrations of starting material were ca. 150 μmol, whereas, kinetic studies were generally conducted at initial concentrations of 15 - 20 μmol.

Sample Analysis. Concentrations were determined using response factors or concentration standard curves. Samples were taken at appropriate time intervals during the reaction and analyzed directly, i.e., no extraction procedures were carried out, by HPLC using one of the following: (1) two Waters Model 6000 pumps equipped with a Waters Model 990 photodiode array detector and accompanying NEC APC-III controller and software using a 0 to 50% acetonitrile/phosphoric acid buffer (pH 2) gradient (Waters Curve No. 8) in 5 min at a flow rate of 2 mL/min on a Waters NOVAPAK 4 μm C-18, 8 mm x 10 cm radial compression module column; (2) a Gilson Model 42 HPLC system equipped with a Gilson Model 116 UV detector (210 and 230 nm monitored), an IBM PC/AT controller and accompanying Gilson software using a gradient of 0 to 5% acetonitrile/phosphoric acid buffer (pH 2) at 2 min, to 20% at 5 min, and to 60% at 7 min (all linear transitions) at a flow rate of 1.5 mL/min on a standard Beckman C-18 (ODS, 5 μm) end-capped 4.6 mm x 25 cm steel jacketed column; or (3) two Waters Model 510 HPLC pumps equipped with a Waters Model 490 UV detector (210, 225 and 235 nm monitored), NEC APC-IV controller and Maxima 820 software using a 15 to 50% acetonitrile/phosphoric acid buffer (pH 2) 10 min gradient (Waters Curve No. 10) at a flow rate of 0.4 mL/min.

Product Isolation. The ozonation reaction was monitored by HPLC as described above and carried out until all starting material was depleted or the maximum concentration of product was obtained. The reaction mixture was extracted with 50 mL ethyl acetate (3X), the extract dried over Na_2SO_4, concentrated in vacuo to ca. 5 mL and then to dryness using nitrogen. The residue was redissolved in several milliliters of acetonitrile and separation of the reaction products achieved by semi-preparative HPLC on the Gilson system with a standard Beckman C-18 (ODS, 5 μm) end-capped 10 mm x 25 cm steel jacketed semi-prep column using a 0 to 20% acetonitrile/water linear gradient in 8 min at a flow rate of 6 mL/min.

Results

Ozonation of Atrazine - Transformation Pathway. Ozonation of atrazine afforded a complex mixture consisting of four primary products [6-amino-2-chloro-4-isopropylamino-*s*-triazine, 6-amino-2-chloro-4-ethylamino-*s*-triazine, 4-acetamido-2-chloro-6-isopropylamino-*s*-triazine and 4-acetamido-2-chloro-6-ethylamino-*s*-triazine] and which were subsequently degraded to three secondary products [2-chloro-4,6-diacetamido-*s*-triazine, 6-amino-4-acetamido-2-chloro-*s*-triazine and 2-chloro-4,6-diamino-*s*-triazine] (Fig. 1 and Table I).

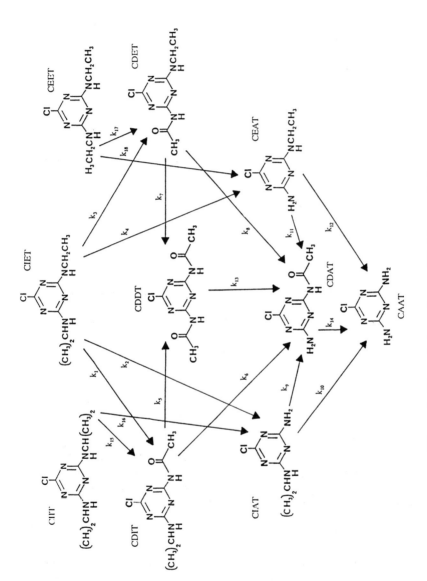

Figure 1. Degradation scheme of 2-chloro-s-triazines.

Structural identification of CAAT, CEAT and CIAT was established by comparison with authentic samples. Syntheses, NMR and mass spectrometry provided structural verification of the remaining products (CDAT, CDDT, CDET and CDIT). Incomplete amidization of CAAT using acetic anhydride afforded a mixture of CAAT, CDAT, and CDDT in ca. 1:2:1 ratio which were separated by HPLC. In a similar fashion CDET and CDIT were synthesized by amidization of CEAT and CIAT, respectively *(18)*.

Table I. 2-Chloro-*s*-triazines and Ozonation Products

Abbreviation	Chemical Name
CAAT	2-chloro-4,6-diamino-*s*-triazine
CDAT	6-amino-4-acetamido-2-chloro-*s*-triazine
CDDT	2-chloro-4,6-diacetamido-*s*-triazine
CDET	4-acetamido-2-chloro-6-ethylamino-*s*-triazine
CDIT	4-acetamido-2-chloro-6-isopropylamino-*s*-triazine
CEAT	6-amino-2-chloro-4-ethylamino-*s*-triazine
CEET	simazine (2-chloro-4,6-diethylamino-*s*-triazine)
CIET	atrazine (2-chloro-4-ethylamino-6-isopropylamino-*s*-triazine)
CIAT	6-amino-2-chloro-4-isopropylamino-*s*-triazine
CIIT	propazine (2-chloro-4,6-diisopropylamino-*s*-triazine)

Ozonation of Atrazine Degradation Products. Treatment of CDIT with ozone yielded CDAT and CDDT initially which then gave rise to formation of CAAT and more CDAT. No CDET, CIAT or CEAT were detected during this reaction. Ozonation of CDET gave rise to CDAT and CDDT which were subsequently transformed to CAAT and CDAT, respectively. No CEAT was formed during this reaction. CEAT and CIAT ozonation afforded CAAT and CDAT. The formation of CEAT was not observed in the ozonation of CIAT.

Prolonged ozonation (> 3 hr) of CAAT eventually gave rise to 6-amino-2-chloro-4-hydroxy-*s*-triazine and nitrate and then to 6-amino-2,4-dihydroxy-*s*-triazine but no 4,6-diamino-2-hydroxy-*s*-triazine as determined by comparison of the HPLC retention times and the UV spectra (200 - 350 nm) with standards. When AAtrex Nine-O (formulated atrazine) was ozonated, a mixture of CDDT, CDAT, and CAAT was formed. Further oxidation beyond CAAT did not occur under similar conditions.

Relative Degradation Rates of Products. The molar concentrations of the atrazine ozonation products as function of time are shown in Figure 2. The initial ratio of $\{[CDIT] + [CIAT]\}$ to $\{[CDET] + [CEAT]\}$ indicates the relative reactivity of the ethyl moiety versus the isopropyl group of atrazine, respectively, and can also be written as $(k_3 + k_4)/(k_1 + k_2)$ (Table II). This ratio decreases with time because the concentrations of CDET and CEAT decrease more rapidly than CDIT and CIAT.

Propazine and Simazine Ozonation. Aqueous ozonation of simazine under neutral conditions, as expected, gave rise to CDET, CEAT, CDDT, CDAT and CAAT. In similar fashion, propazine ozonation yielded CDIT, CIAT, CDDT, CDAT and CAAT (unpublished results). A solution containing the above two s-triazine pesticides was ozonated to determine the relative reactivities. First order rate constants were obtained after equilibrium was reached (ca. 50 min) (Fig. 3). The ratio of the disappearance rates of CIIT to CEET can be written as $(k_{15} + k_{16})/(k_{17} + k_{18})$, which represents the relative reactivity of the isopropyl and ethyl moieties, and was determined to be 0.206.

Competitive ozonation of CIAT and CEAT. Ozonation of a solution containing CIAT and CEAT was carried out and first order rate constants obtained after equilibrium was reached (ca. 50 min) (Fig. 4). The ratio of the disappearance rate of CIAT to CEAT, which can also be described as $(k_9 + k_{10})/(k_{11} + k_{12})$, was determined to be 0.200 and also reflects the relative reactivity of the isopropyl and ethyl groups.

Table II. Reaction of Ethyl Versus Isopropyl Moieties of Atrazine

Time (min)	[CEAT] + [CDET] μmol/L	[CIAT] + [CDIT] μmol/L	$\dfrac{k_3 + k_4}{k_1 + k_2}$
10	2.4	11.2	0.214
30	4.1	20.7	0.198
80	4.7	37.5	0.125
180	3.7	50.8	0.073

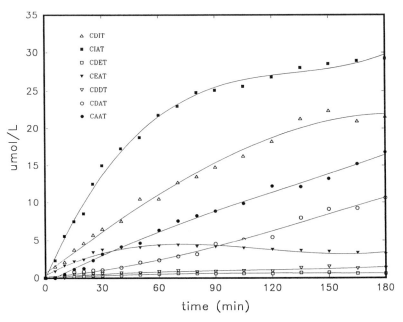

Figure 2. Atrazine reaction profile (formation of ozonation products; [atrazine]$_0$ = 153 μmol).

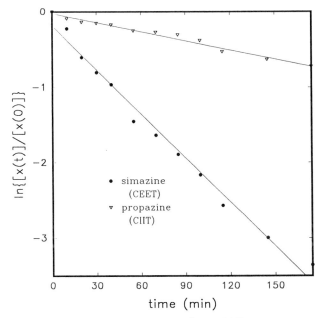

Figure 3. Degradation of simazine (r^2 = 0.985) versus propazine (r^2 = 0.978).

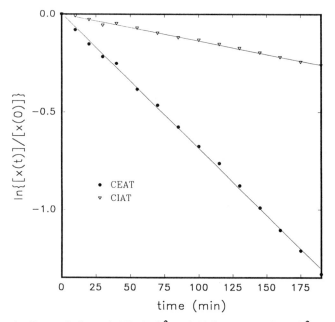

Figure 4. Degradation of CEAT (r^2 = 0.999) versus CIAT (r^2 = 0.996).

Discussion and Conclusions

Analysis of the aqueous ozonation of atrazine and its degradation products demonstrated that amino alkyl groups are the first site of attack. The N-alkyl group is either removed or converted to the N-acetyl and the s-triazine ring remains intact. Furthermore, the isopropyl group is not converted to the ethyl group nor are the ethyl or isopropyl moieties converted to an aldehyde *(18)*. The ozonation reactions of CDET and CDIT clearly demonstrate that the N-alkyl group is far more reactive than the N-acetyl moiety. Additionally, oxidation of the amino group was shown to occur only after prolonged ozonation times giving rise to nitrate and the corresponding hydroxy-s-triazine. The chlorine was not removed until it was made somewhat more reactive as it is with 6-amino-2-chloro-4-hydroxy-s-triazine.

Previous ozonation mechanisms of amines in organic solvents cannot explain the formation of CDET from atrazine and simazine or the formation of CDAT from CEAT, i.e., formation of the acetyl moiety from the ethyl group is precluded *(19, 20)*. Furthermore, ozone attack was favored at the tertiary hydrogen as demonstrated when diisopropylethylamine was used as a substrate *(19)*. In the current study, however, reaction with the ethyl moiety was overwhelmingly preferred over the isopropyl. More likely, then, the ozonation of 2-chloro-N-alkylatedamino-s-triazines in water does not proceed via direct reaction with ozone, rather the reaction probably involves a hydroxy radical mechanism.

If hydroxy radical attack occurred at the primary methyl, then the isopropyl group would be expected to react at a rate twice that of the ethyl moiety. This was not the case. Alternatively, the carbon-hydrogen bond of the isopropyl group, H-$C(CH_3)_2$, is a slightly weaker bond (95 kcal/mol) than the corresponding ethyl hydrogen (98 kcal/mol) and the C-H bond of the methyl group is stronger still (104 kcal/mol) *(21)*. Furthermore, if the reaction is electronically controlled, then hydroxy radical attack on the hydrogen of the more electron rich carbon (the hydrogen on the carbon α to the nitrogen) would be favored with the isopropyl hydrogen favored slightly over the ethyl. Yet, if the reaction is actually sterically controlled, then the more hindered isopropyl hydrogen would be less reactive. The results conclusively showed that reaction was preferred 5:1 on the ethyl over the isopropyl moiety suggesting, therefore, that attack occurs at the electron rich α-carbon and that the reaction is sterically controlled.

In summary, these results have demonstrated the fate of the s-triazine moiety and the alkyl groups and will provide some of the necessary details to proceed in the optimization of s-triazine remediation. Investigations are continuing to discern the extent of hydroxy radical involvement and the role of other oxidizing species in the ozonation of atrazine, and to quantitate the preference of oxidation and/or removal of the ethyl, isopropyl and acetyl nitrogen substituents. With the appearance of s-triazine residues not only in agricultural sites but in ground and surface waters as well, these results will certainly be useful to waste remediation investigators, and may also be helpful to those developing methods for triazine residue removal in water treatment.

Acknowledgement

The author expresses her sincere appreciation to Julie C. Lin and Joseph J. O'Connell for their technical assistance. This work was funded in part by a Cooperative agreement between the U.S. Department of Agriculture and CIBA-GEIGY Corporation.

Disclaimer

Mention of specific products or suppliers is for identification and does not imply endorsement by the U.S. Department of Agriculture to the exclusion of other suitable products or suppliers.

Literature Cited

1. Somich, C.J.; Muldoon, M.T.; Kearney, P.C. "On-Site Treatment of Pesticide Waste and Rinsate Using Ozone and Biologically Active Soil". *Environ. Sci. Technol.* **1990**, *24*, 745-749.
2. Aharonson, N. "Potential Contamination of Groundwater by Pesticides". *Pure Appl. Chem.* **1987**, *59*, 1419-1446.
3. Parsons, D.W.; Witt, J.M. "Pesticides in Groundwater in the United States of America. A Report of a 1988 Survey of State Lead Agencies." **1988**. Oregon State University Extension Service, Corvallis, Oregon.
4. Kearney, P.C.; Muldoon, M.T.; Somich, C.J.; Ruth, J.M.; Voaden, D. J. "Biodegradation of Ozonated Atrazine as a Wastewater Disposal System". *J. Agric. Food Chem.* **1988**, *36*, 1301-1306.
5. Leeson, A.; Hapeman, C.J.; Shelton, D.R. "Biomineralization of Atrazine Ozonation Products. Application to the Development of a Pesticide Waste Disposal System." *J. Agric. Food Chem.* **1993**, *41*, 983-987.
6. Glaze, W.H. "Drinking-Water Treatment with Ozone". *Environ. Sci. Technol.*, **1987**, *21*, 224-230.
7. Hoigne, J. "The Chemistry of Ozone in Water". In *Process Technologies for Water Treatment*; Stucki, S., Ed.; Plenum Publishing Corp.: New York, 1988; pp 121-143.
8. Hoigne, J.; Bader, H. "Rate Constants of Reactions of Ozone with Organic and Inorganic Compounds in Water - I. Non-Dissociating Organic Compounds". *Water Res.* **1983**, *17*, 173-183.
9. Hoigne, J.; Bader, H. "Rate Constants of Reactions of Ozone with Organic and Inorganic Compounds in Water - II. Dissociating Organic Compounds". *Water Res.* **1983**, *17*, 184-194.
10. Glaze, W.H. "Reaction Products of Ozone: A Review". *Environ. Health Perspectives* **1986**, *69*, 151-157.
11. Peyton, G.R.; Gee, C.S.; Smith, M.A.; Brady, J.; Maloney, S.W. "By-Products from Ozonation and Photolytic Ozonation of Organic Pollutants in Water: Preliminary Observations." In *Biohazards of*

Drinking Water Treatment; Larson, R. A., Ed.; Lewis Publishers: Chelsea, Michigan, 1989; pp 185-200.

12. Glaze, W.H.; Kang, J.-W. "Advanced Oxidation Processes for Treating Groundwater Contaminated with TCE and PCE: Laboratory Studies". *J. Am. Water Works Assoc.* **1988**, *80*, 57-63.

13. Peyton, G.R.; Glaze, W.H. "Mechanism of Photolytic Ozonation". *ACS Symp. Ser.* **1987**, *No. 327*, 76-88.

14. Staehelin, J.; Hoigne, J. "Decomposition of Ozone in Water in the Presence of Organic Solutes Acting as Promoters and Inhibitors of Radical Chain Reactions". *Environ. Sci. Technol.* **1985**, *19*, 1206-1213.

15. Staehelin, J.; Hoigne, J. "Decomposition of Ozone in Water: Rate of Initiation by Hydroxide Ions and Hydrogen Peroxide". *Environ. Sci. Technol.* **1982**, *16*, 676-681.

16. Cook, A.M. "Biodegradation of *s*-Triazine Xenobiotics". *FEMS Microbiol. Rev.* **1987**, *46*, 93-116.

17. Hapeman-Somich, C.J.; Zong, G.-M.; Lusby, W.R.; Muldoon, M.T.; Waters, R. "Aqueous Ozonation of Atrazine. Product Identification and Description of the Degradation Pathway." *J. Agric. Food Chem.* **1992**, *40*, 2294-2298.

18. Somich, C.J.; Kearney, P.C.; Muldoon, M.T.; Elsasser, S. "Enhanced Soil Degradation of Alachlor by Treatment with Ultraviolet Light and Ozone". *J. Agric. Food Chem.*, **1988**, *36*, 1322-1326.

19. Bailey, P.S.; Southwick, L.M.; Carter, T.P., Jr. J. Org. Chem. **1978**, *43*, 2657-2652.

20. Bailey, P.S. "Ozonation in Organic Chemistry. Vol. II. Nonolefinic Compounds." **1982**. Academic Press, New York, New York.

21. Golden, D.M.; Benson, S.W. *Chem. Rev.* **1969**, *69*, 125.

RECEIVED October 27, 1993

RADIOACTIVE AND MIXED-WASTE MANAGEMENT

Chapter 15

Cyclic Gas Releases in Hanford Site Nuclear Waste Tanks

D. A. Reynolds and H. Babad

Westinghouse Hanford Company, P.O. Box 1970, MSIN R2-11,
Richland, WA 99352

At the Hanford Site, 177 tanks have been constructed for storing radioactive waste; 53 tanks have been identified as having potential safety issues. This report focuses on tank 241-SY-101, which generates hydrogen gas and releases this gas on a cyclic schedule. The tank is extensively instrumented to monitor the hydrogen gas, and work restrictions minimize the risk of a hydrogen burn. Mitigation testing centers on two in-tank test assemblies: (1) a large mixer pump; and (2) dilution, heating, and ultrasonic devices. The current understanding of what is happening in the tank is presented in this paper.

Background. The Hanford Site, located in southeast Washington State (Figure 1), comprises about 900 km² (560 mi²) of land. The Hanford Site was part of the Manhattan Project, which started in 1943. As part of this defense effort, towns, reactors, chemical separation plants, and waste storage tanks were built in a remarkably short period of time; in fact, the Manhattan Project was a larger construction effort than the Panama Canal. Within 27 months from start of construction, the Hanford Site produced the explosive charge for the first nuclear detonation.

National defense needs for plutonium kept the Hanford Site growing. Eventually, nine plutonium production reactors were built, and additional chemical separation plants were built as better technology came along. Waste management efforts also grew with increased production.

As national defense needs for plutonium were met, the production reactors were shut down. The last plutonium reactor on the Hanford Site was placed in cold standby in 1988. The Hanford Site mission has changed from plutonium production

0097–6156/94/0554–0236$08.00/0

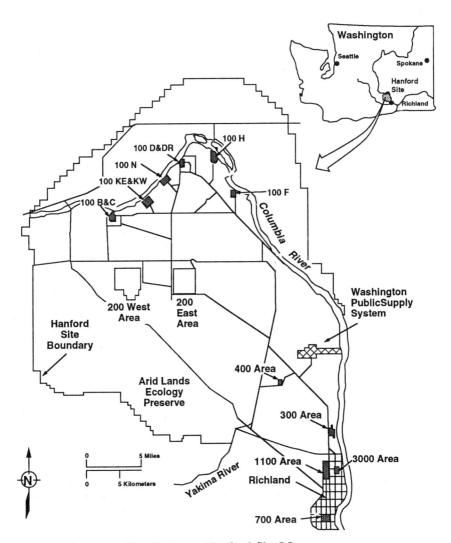

Figure 1. Hanford Site Map.

to environmental restoration. Much of this activity centers around the disposal of the approximately 227 ML (60 Mgal) of radioactive waste that is stored in 177 tanks.

Tank Design and Capacity. Originally waste tanks were designed with a single shell: a concrete tank with a mild steel liner. The single-shell tanks were 22.9 m (75 ft) in diameter and varied in height. Three different single-shell tank designs had nominal capacities of 1.9, 2.8, and 3.8 ML (500,000 gal, 750,000 gal, and 1,000,000 gal). Several 190,000 L (50,000 gal) tanks were also constructed. Between 1943 and 1965, 149 single-shell tanks were constructed. These tanks currently contain 140 ML (37 Mgal) of waste. In 1980, all single-shell tanks were pumped of easily pumped liquid and placed on inactive status. Stabilizing and isolating efforts for these tanks have continued since that time.

Leaks in the older single-shell tanks prompted a design change, and the first double-shell tank farm was constructed in 1965. The double-shell tank is a 22.9-m-dia. (75-ft-dia.) concrete tank with two mild steel liners separated by an annulus (Figure 2). Leak detection devices are in the primary tank (surface level), the annulus, and under the tank. None of the double-shell tanks have leaked. Since 1965, 28 double-shell tanks have been built at the Hanford Site and all remain active in the waste management program. The double-shell tanks receive waste from plants and customers and serve as feed tanks to evaporators and disposal plants.

Safety assessment reports are written for nuclear facilities in accordance with U.S. Department of Energy (DOE) policy. In 1989, certain safety questions came up that were beyond the scope of the safety assessments then in use. The concerns centered around tanks with high heat, high fuel content (either organic or ferrocyanide), and tanks that produced flammable gas. Fifty-three tanks of concern were identified, categorized into four different watch lists, and placed under operating restrictions. Tanks can be listed in more than one category.

Twenty-three tanks (5 double-shell tanks and 18 single-shell tanks) have the potential for accumulating hydrogen or other flammable gases above the flammability limit (Table I). Only the five double-shell tanks are now showing signs of actively releasing hydrogen. Tank 241-SY-101 (101-SY), the focus of this report, is the tank of greatest concern.

Cyclic Gas Release Events at Tank 101-SY

Tank 101-SY was filled between 1976 and 1980 with two different types of wastes: double-shell slurry and complexant concentrate. Complexant concentrate is waste that has a relatively high level of water-soluble organic complexants. Double-shell slurry is the most concentrated product that the evaporators can make. From the first addition of waste, tank 101-SY exhibited what is known as slurry growth. Slurry growth is an increase in surface level without a corresponding addition of waste. Shortly after the last waste was put into the tank in 1980, the slurry growth was followed by a sudden drop in surface level. The cycle of slurry growth followed by a surface level drop has been occurring on approximately 100-day intervals since that time.

Table I. Flammable Gas Tanks

Flammable Gas Tanks	Maximum Temperature
101-A	69 °C (156 °F)
101-AX	61 °C (142 °F)
103-AX	46 °C (114 °F)
102-S[1]	43 °C (110 °F)
111-S	35 °C (95 °F)
112-S	30 °C (86 °F)
101-SX	61 °C (142 °F)
102-SX	70 °C (158 °F)
103-SX	83 °C (181 °F)
104-SX	79 °C (174 °F)
105-SX	87 °C (188 °F)
106-SX[1]	47 °C (116 °F)
109-SX[2]	64 °C (148 °F)
110-T	17 °C (63 °F)
103-U	33 °C (91 °F)
105-U	33 °C (92 °F)
108-U	33 °C (91 °F)
109-U	33 °C (91 °F)
103-AN[3]	43 °C (110 °F)
104-AN[3]	43 °C (110 °F)
105-AN[3]	40 °C (104 °F)
101-SY[3]	58 °C (136 °F)
103-SY[3]	47 °C (117 °F)

[1]Tank also on Organic Watch List.
[2]Tank has potential for flammable gas accumulation only, because other SX tanks vent through it.
[3]Double-shell tank.

Figure 3 shows the surface level of tank 101-SY over a 4-yr period; the cyclic rise and fall of the tank's contents is shown clearly. The time between major falls is a nominal 100 days, and drops are typically 13 to 25 cm (5 to 10 in.).

Double-Shell Tanks

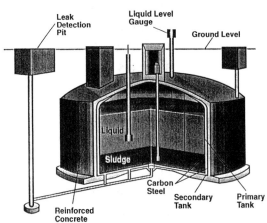

- 28 Tanks Constructed Between 1968-86

- 3,785 m^3 to 4,315 m^3 (1 to 1.14 Mgal) Capacity

- Tanks Currently Contain

  ~ 9 x 10^4 m^3 (24 Mgal) of Mostly Liquids (Also Sludges and Salts)

  ~ 400 x 10^{16} Bq (110 MCi)

- None Have Leaked

Figure 2. Typical Double-Shell Tank.

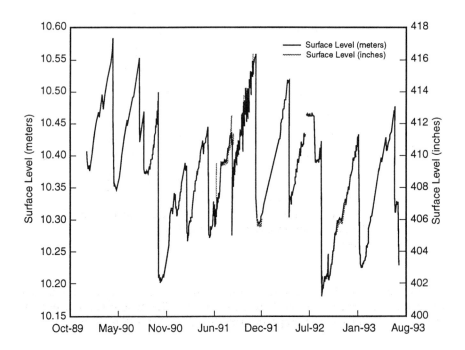

Figure 3. Surface Level of Tank 101-SY.

The double-shell tanks have a ventilation system that keeps the tank dome space at a slight vacuum. Tank 101-SY is kept at about -750 Pa gauge (-3 in. water gauge). Initial observations noted that the drops in surface level corresponded with an increase in pressure. Occasionally, the tank pressurizes for a short period of time, which is indicative of a gas release.

A thermocouple tree in the tank has thermocouples positioned every 0.61 m (2 ft). Figure 4, based on data gathered from the thermocouple tree, shows a typical temperature profile before and after a drop. Before the drop, there is a parabolic temperature profile at the bottom of the tank; this is expected from a self-heating solid. The heat in this tank is due to radiodecay of ^{137}Cs. Above this layer there is a flat temperature profile. A flat profile is typical of a convective heat transfer zone when there is a hot layer below it. Consequently, the two layers are termed as the nonconvective layer and the convective layer. Figure 4 shows that the temperature profile straightens considerably after a drop with hotter material rising to near the surfaces, indicating that the tank became well mixed.

Figure 5 shows a schematic of what is happening in tank 101-SY. Solids in the tank settle to the bottom of the tank and form a nonfluid layer. The gas generated in the waste is trapped in this layer. When enough gas is trapped in the nonconvective layer, buoyancy forces overcome the static forces and the tank contents roll over. The tank contents mix and 200 to 300 m^3 (7000 to 10,000 ft^3) of gas are released to the tank dome during the gas release event. The solids fall to the bottom and the cycle begins again.

The safety concern derives from the gas composition. Gas samples from the ventilation system were analyzed using a mass spectroscope or graph. The ventilation system sample is primarily air with small portions of gas. The best estimate is the composition of the gas coming from the waste, or else generated in the waste, as shown in Figure 6. The gas exists in the waste tank as a flammable mixture; but, because of the water and small bubbles in the waste, it is not expected to be a hazard in the tank. However, concern arises when the gas is mixed with air. When this occurs, the hydrogen concentration has exceeded 5 percent for short periods of time, and this exceeds the lower flammable limit.

Work restrictions have been put in place to prevent such things as sparks and heat sources in the tank and surrounding area. The tank has had additional instruments installed, and surveillance activities have increased. Currently, there are three electrochemical hydrogen cells, a metal oxide semiconductor hydrogen detector, a thermal conductivity hydrogen meter, a gas chromatograph, an on-line mass spectrometer, and flow and pressure meters. Additional instruments are planned to be installed for various purposes.

Data from a typical gas release event are shown Figures 7 and 8. The pressure increase and flow increase decay away rapidly. Both pressure and flow have reached normal ranges by the time the hydrogen concentration has increased in the vent header. The surface level drop has been abrupt, but will continue down for several days before starting to increase. The temperature profile shows the warmer material from the bottom moves to the top of the tank. A different view shows the temperature at location 4, which is the maximum temperature before the gas release

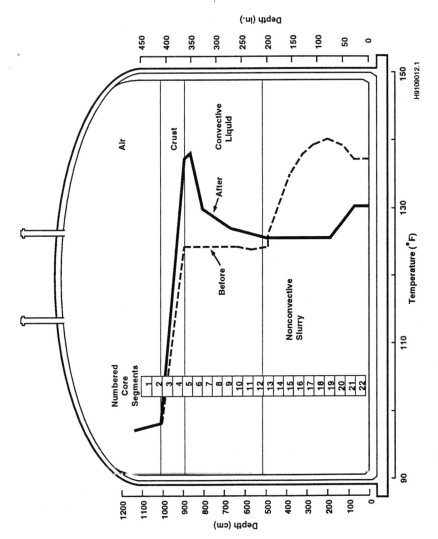

Figure 4.   Temperature Profile of Tank 101-SY.

Proposed Mechanism of Tank 101-SY Venting

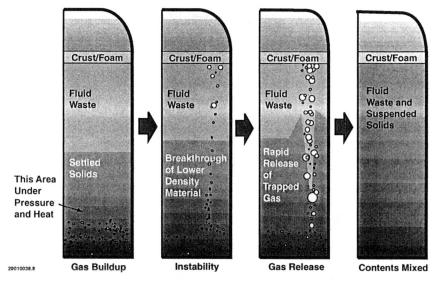

Figure 5. Schematic of Activity Inside Tank 101-SY.

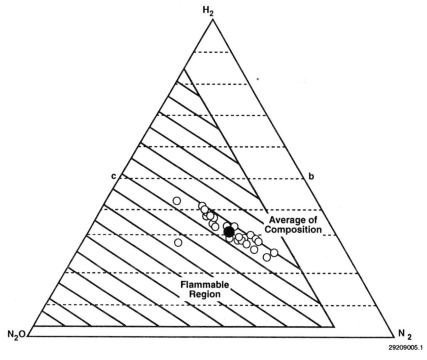

Figure 6. Estimated Gas Composition in Tank 101-SY Waste.

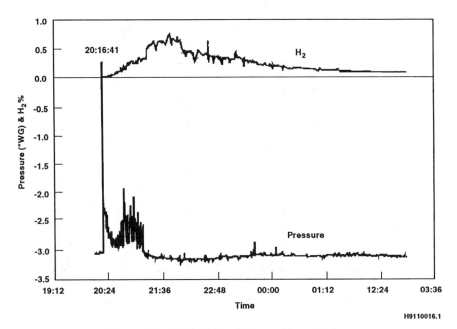

H9110016.1

Figure 7. Typical Gas Release Event Data.

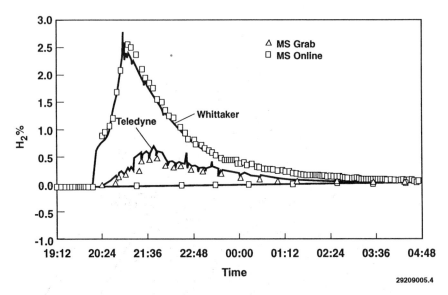

29209005.4

Figure 8. Hydrogen Release Profile, May 1991 Gas Release Event.

event, and location 16, which is just below the surface. These curves show that the temperature changes in about 1.5 minutes.

Waste Composition

Two core samples have been taken from tank 101-SY; twenty-two segments [each 28 cm (19 in.) long] extend from the surface to the bottom of the tank. These segments verify that there are two layers in the tank. Chemically there are few differences between the two layers, but physically there is a great difference.

The primary chemicals are sodium salts of nitrate, nitrite, hydroxide, and aluminate. Various other chemicals are present in smaller amounts. Table II shows the chemical composition of the waste.

Table II. Chemical Composition of the Waste
(All Values in Weight Percent)

Component	Convective	Nonconvective
Water	36.0	32.0
NO_3^-	10.3	10.2
NO_2^-	11.7	9.1
TOC	1.25	1.9
Na^+	20.1	21.2
Al	3.4	3.5
Ca^{+2}	0.012	0.036
Cr	0.088	0.7

TOC = total organic carbon

The major radioactive species is ^{137}Cs, a specie with a 30-yr half-life. Strontium-90 has a 29-yr half-life, but is present is smaller quantities. The transuranic elements, plutonium and americium, are present in large enough quantities to classify this waste as transuranic waste.

The density of the waste shows a difference between the two layers. The liquid convective layer has a density of about 1.5, while the nonconvective layer has a density of about 1.7. The amount of solids is also much greater in the nonconvective layer. The low viscosity of the liquid allows the convective layer to move easily.

One of the types of waste that was put into tank 101-SY was complexant concentrate. This waste had large amounts of ethylenediametetraacetic acid (EDTA), hydroxy EDTA (HEDTA), and citric acid when the waste was produced. The waste is between 1 and 2 weight percent total organic carbon, but very few of the starting complexants are present. Most of the organics are most likely degradation products of the EDTA and HEDTA starting compounds. However, the total identified complexant fragments represented only about 12 percent of the total organics in the waste. The rest of the fragments are water-soluble organics believed to be mostly sodium formate and sodium oxalate, the end product of complexant degradation.

Steps Taken to Understand the Problem

Measurements of the tank parameters and core samples gave a good basis for understanding the tank contents and mechanisms related to the rollover. However, a good understanding of the basic gas-generating mechanism was not available.

Initially, hydrogen gas was thought to be generated by radiation. Simulated waste experiments in the mid-1980s showed that hydrogen gas also could be generated without radiation. The simulated waste experiments also generated nitrous oxide and nitrogen. The kinetics seemed to indicate that chemical reactions could account for the total gas production, but the chemical reactions remained unknown.

In 1990, studies were started at Argonne National Laboratories (ANL), Georgia Institute of Technology (GIT), and Pacific Northwest Laboratories (PNL) to obtain better details of important reactions. Argonne focused on radiation reactions, and GIT looked at the organic chemical reactions. PNL studied how the gas is retained in the waste.

Argonne established that the nitrate in the tank contents suppressed the radiolysis of water. The radiolysis of water is not a viable mechanism for producing the amount of hydrogen found in tank 101-SY. Argonne also established the following:

- The organics produced hydrogen in proportion to the amount of organic present.

- Radiation plays a role in producing an intermediate specie.

- Radiated solution produced more gas than nonirradiated solution, even out of the radiation field.

GIT showed that the starting complexants can degrade, with the hydroxide and nitrate, to simpler molecules. Eventually, the formate ion is formed, which can react via the Cannizzaro Reaction to produce hydrogen. GIT also established the following:

- The nitrous oxide and nitrogen produced come from the nitrite.

- Ammonia is produced from the amine groups in the complexants.

The studies at PNL showed that the tank's high salt content suppressed the solubility of nitrous oxide, and that surface tension may be holding the gas bubbles onto solid particles. Occasionally bubbles have been seen inside of solid particle clumps. PNL also studied corrosion and iron catalyzed reactions.

The laboratory tests have added greatly to our understanding of the radiochemical reaction taking place in the tank 101-SY. Further work will be done in the coming year to further this understanding.

Steps Taken to Mitigate the Problem

While the probability of a gas ignition event is extremely low, the level of risk from such an event remains unacceptable because of potentially high consequences to human health and the environment. Therefore, after initial but detailed characterization of the contents of tank 101-SY (and other tanks that undergo cyclic venting), steps to mitigate the problem with the flammable tanks will be evaluated and ultimately implemented.

The problem in tank 101-SY is not the total amount of hydrogen produced; instead, it is the potential of all the hydrogen being released at once. If the gas were released as generated, the hydrogen concentrations would remain below a level of concern.

Mitigation will not prevent or stop the radiochemical reactions, but it will bring the tank under control. The hydrogen will be released either continuously or frequently, as dictated by the operations. Once that is done, the problem can be controlled safely until final disposal of the waste.

Potential mitigation methods, all aimed at minimizing or eliminating cycles, include heating and/or dilution of the wastes and/or stirring them to allow the gases that are formed by the chemicals and radionuclides to vent continuously. Another alternative being explored is ultrasonics for forcing continuous release of the gases as formed. These alternatives are not mutually exclusive, because both pumping and use of ultrasound generate heat in the tank. A two-phase mitigation strategy for tank 101-SY has been initiated. After laboratory-scale data collection, a large-scale test of the four potential methods will be run in tank 101-SY. A pumping test will be done with a variable-speed liquid jet pump placed directly in the waste.

The remaining alternatives will be tested in a test chamber that contains apparatus for measuring the effects of heating and dilution and/or treatment of the waste with ultrasound. This steel test chamber will isolate the test region from the rest of the tank, and will provide a more controlled environment for evaluating these mitigation techniques.

The first test will involve mixing. The theory is that the natural rollover mixes the waste by bringing the gas-laden material from the bottom of the tank to the surface, where the gas can be released. Mixing will simulate the natural phenomena.

The mixing test will be accomplished by putting a jet mixer pump into a riser that is off-center of the tank. The 150-horsepower pump is designed to take liquid from the convective zone and jet it into the nonconvective zone near the bottom of the tank. Two opposing jets will be used, and there are provisions for rotating the jets to sweep the entire tank. The jets will sluice the material from the nonconvective zone, and will eventually dislodge and mix the solids.

The mixing will begin at a very low rate of speed until the magnitude of the jet effects are known. The low speed will prevent the perturbation of a large gob of waste, which could rise to the surface and release more hydrogen gas than expected. The speed of the pump will be increased slowly and the jets will be rotated.

The surface level and the hydrogen gas in the vapor space will be monitored closely. These readings are expected to provide the best understanding of the effect of the mitigation test.

Actual waste shows that shear strength and viscosity are dependent on temperature. By heating the tank modestly to about 75 °C the crystals could redissolve, and the nonconvective layer could lose strength. If these events occur, gas bubbles would be able to rise through the weakened matrix and escape.

The test chamber will have coil heaters that will heat a captured portion of the nonconvective layer. The enclosed headspace will be monitored closely for hydrogen. An increase of hydrogen will indicate that hydrogen is escaping from the waste due to heating.

Tanks that hold and episodically release flammable gas all contain concentrated waste. The more dilute waste tanks do not exhibit these phenomena. Therefore, dilution of the concentrated waste could redissolve solids, decrease viscosity, and slow chemical reactions. Only modest dilution is possible in tank 101-SY, however. A 30 percent dilution is about the most that can be accommodated in the existing tank farm.

The test chamber will have means of introducing water, dilute hydroxide, or 3 molar hydroxide solutions to the waste to see if a 30 percent dilution will aid the release of gas. Chemical factors important to the success of this technique would be the behavior of the aluminum salts during the dilution process.

Ultrasound has been used successfully in other industries for controlling foam and causing gas bubbles to disengage. Some laboratory work showed that, on a small scale, bubbles could be concentrated in zones. An ultrasonic device has been designed into the chamber to test this method of mitigation. One of the difficulties with ultrasonic devices, however, is that energy is adsorbed in a very short distance by the solid particles. The attenuation factors of actual waste will be measured in the test chamber.

Conclusions

The hydrogen generation in the aqueous waste at the Hanford Site has posed some difficult scientific problems. The understanding of what is happening in the tank has increased greatly over the past two years. The controls placed on the operations in the tank farm have been successful in preventing any accidents. Steps are being taken rapidly to control and mitigate the hydrogen release cycle.

RECEIVED January 14, 1994

Chapter 16

Mechanistic Studies Related to the Thermal Chemistry of Simulated Nuclear Wastes That Mimic the Contents of a Hanford Site Double-Shell Tank

E. C. Ashby, E. Kent Barefield, Charles L. Liotta, H. M. Neumann, Fabio Doctorovich, A. Konda, K. Zhang, Jeff Hurley, David Boatright, Allen Annis, Gary Pansino, Myra Dawson, and Milene Juliao

School of Chemistry and Biochemistry, Georgia Institute of Technology, Atlanta, GA 30332

Certain aspects of the thermal reactions of organic complexants in simulated aqueous nuclear waste mixtures containing NaOH, $NaNO_2$, $NaNO_3$ and $NaAl(OH)_4$ have been investigated. Since formaldehyde is expected as a degradation product from various complexants, the factors affecting the reaction by which H_2 gas is formed from formaldehyde and OH^- have been studied. A mechanism consistent with these results is presented. The source of the nitrogen atoms in the gases N_2O, N_2, and NH_3 formed from the degradation of HEDTA has been determined using ^{15}N labeling experiments. It was found that the nitrogens in N_2O and N_2 are produced from $NaNO_2$ whereas 88% of the nitrogen in NH_3 come from $NaNO_2$ and 12% come from HEDTA. A number of degradation products formed in series in the long-term thermal degradation of HEDTA at 120° C were identified by ^{13}C NMR. In the reaction of glycolate in simulated waste, formate and oxalate are formed, and nitrite (but not nitrate) decreases in amount. Kinetic studies involving glycolate decomposition have resulted in the determination that the reaction is first order in each of the reactants: glycolate, nitrite, and aluminate. A mechanism is proposed for decomposition of glycolate. A scheme for the sequential degradation of HEDTA, based on the same mechanistic ideas, is also presented. Finally, some observations on the effect that O_2 has on the decomposition of organics and on the gas generation in simulated waste are given.

Defense nuclear wastes are currently stored by DOE at Hanford, Washington and managed by the Westinghouse Hanford Co. At this site are 177 storage tanks: 149 single shell and 28 double shell tanks. Of these tanks, 23 are suspected of flammable gas evolution; 5 of these are double shell tanks and 18 are single shell tanks. One of the double shell tanks, labeled 241-SY-101 and which is referred to as Tank 101-SY,

0097–6156/94/0554–0249$10.88/0

is the most troublesome. Tank 101-SY releases, in events approximately 100 days apart, usually 7,000-11,000 cu. ft. of gas containing H_2, N_2O, and N_2. Other gases such as NH_3 and CH_4 are also released in smaller amounts. The flammability and potential explosive nature of this gaseous mixture is cause for concern and a great deal of effort is now in motion to deal with this problem.

It is clear that before one can arrive at a proposed solution to the episodic release of gases from storage tanks that contain high level nuclear waste, it would be advantageous to determine the chemical origins of the gases and the corresponding mechanistic pathways whereby these gases are produced. Understanding the chemistry involved in tank 101-SY, which contains a large number of organic and inorganic compounds and radionuclides reacting under thermal and radiolytic conditions, presents a formidable problem. In this connection, several laboratories (Pacific Northwest Laboratories (PNL), Richland, WA; Argonne National Laboratories (ANL), Argonne, IL; Westinghouse Hanford (WHC), Hanford, WA and Georgia Institute of Technology (GIT), Atlanta, GA), have been involved in a joint effort to understand the chemistry involved in Tank 101-SY. The ANL organization has concentrated on studying the radiation chemistry related to that in Tank 101-SY, whereas PNL and GIT have concentrated on studying the corresponding thermal chemistry. The ANL, PNL, and GIT studies have involved a simulated waste which mimics the contents in Tank 101-SY. The WHC organization has concentrated on studying the composition of both the volatile and non volatile products obtained from Tank 101-SY. The WHC studies along with some studies conducted at PNL have involved actual core samples from Tank 101-SY.

Two studies carried out previous to the present study were particularly informative. The first study published in 1980 by Delegard (1) provides information concerning the thermal behavior of simulated waste. Specifically kinetic studies dealing with the thermal decomposition of HEDTA and other organic complexants in the presence of $NaAlO_2$, $NaNO_2$, $NaNO_3$, $NaOH$, and Na_2CO_3 were reported as well as identification of the gases evolved in the reaction of HEDTA with this simulated waste. These studies led to a proposed reaction sequence for the decomposition of HEDTA. The second report was published in 1986 by workers at PNL (2). The study was directed to determining the organic decomposition products in Tank 107-AN, a tank to which organic complexing agents had been added. Over 75% of all organic products were identified.

All of our studies at GIT were carried out at 120°C on a simulated waste that mimics the known composition of Tank 101-SY with respect to $NaNO_2$, $NaNO_3$, $NaAlO_2$, and Na_2CO_3. In addition, one or more organic complexants were added. All reactions were strictly thermal (no radiation producing materials present). The high temperature (120° C.) was necessary in order that reactions could be studied in a reasonable period of time. For example, the extent of HEDTA reaction is about 70% after 1,500 hours (~ 62 days) in simulated waste at 120° C, whereas the temperature in 101-SY is closer to 60° C. Reactions at 60° C would take much too long to provide a practical study.

The composition of the simulated waste (termed SY1-SIM-91B) used in this study is presented in Table I.

Table I. Composition of Simulated Waste (SY1-SIM-91B) used in this study

Component	Molarity
Organic	0.21
NaNO$_2$	2.24
NaNO$_3$	2.59
NaAlO$_2$	1.54
Na$_2$CO$_3$	0.42
NaOH	2.00

We have used a number of names and abbreviations for structures in this paper, that may not be familar to all readers, therefore we wish to define these names and abbreviations with the proper chemical formulas.

EDTA -   (HO$_2$CCH$_2$)$_2$NCH$_2$CH$_2$N(CH$_2$CO$_2$H)$_2$
HEDTA -   HOCH$_2$CH$_2$N(CH$_2$CO$_2$H)CH$_2$CH$_2$N(CH$_2$CO$_2$H)$_2$
Glycolic Acid -   HOCH$_2$CO$_2$H
Glycolate -   HOCH$_2$CO$_2^-$
Formate -   HCO$_2^-$ (or HCOO$^-$)
Oxalate -   C$_2$O$_4^=$
EA -   H$_2$NCH$_2$CH$_2$OH
Glycine -   H$_2$NCH$_2$CO$_2$H
ED3A -   HO$_2$CCH$_2$N(H)CH$_2$CH$_2$N(CH$_2$CO$_2$H)$_2$
U-EDDA -   H$_2$NCH$_2$CH$_2$N(CH$_2$CO$_2$H)$_2$
S-EDDA -   HO$_2$CCH$_2$N(H)CH$_2$CH$_2$NHCH$_2$CO$_2$H
EAMA -   HOCH$_2$CH$_2$NHCH$_2$CO$_2$H
IDA -   HN(CH$_2$CO$_2$H)$_2$

ED3A Lactam -

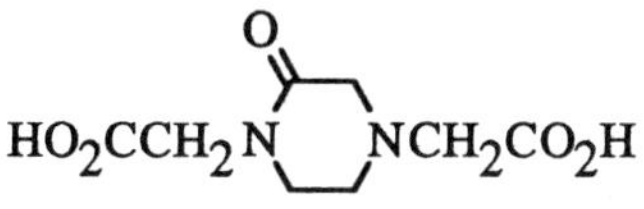

HO$_2$CCH$_2$N   NCH$_2$CO$_2$H

NTA -   N(CH$_2$CO$_2$H)$_3$

In this paper answers to the following questions are pursued:
What is the source of H$_2$?
What is the source of N$_2$O?
What is the source of N$_2$?
What is the source of NH$_3$?
What is the mechanism of degradation of HEDTA and glycolate?
The current status of investigations motivated by these questions will be presented with supporting experimental data.

Source of Hydrogen (27)

A mechanistic explanation for the formation of H_2 in nuclear waste tanks has not been forthcoming except for the H_2 that is produced by the radiolysis of water. It is known that under radiolytic conditions, solutions of EDTA (3), NTA (4), and IDA (5) decompose to formaldehyde. As a consequence, it is reasonable to assume that formaldehyde is formed in nuclear waste tanks containing these and similar complexing agents. It is also known that formaldehyde reacts in base to form equimolar amounts of formate and methanol (Equation 1).

$$2 \ \overset{\overset{\displaystyle O}{\|}}{HCH} \ \xrightarrow{\ OH^-\ } \ \overset{\overset{\displaystyle O}{\|}}{HCO^-} \ + \ CH_3OH \qquad (1)$$

This is known as the Cannizzaro reaction and it proceeds in high yield for relatively high concentrations of formaldehyde. This reaction has been presumed to proceed through an intermediate (3) which functions as a hydride ion donor toward formaldehyde (Scheme 1) (6). On the other hand, it seems reasonable to assume that

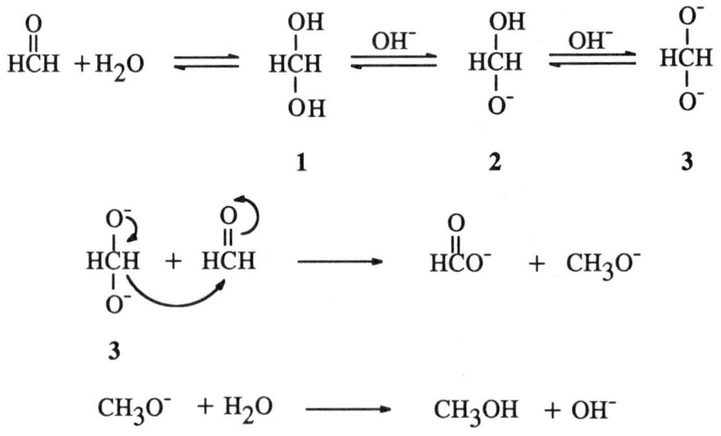

Scheme 1. Mechanism of the Cannizzaro Reaction.

the Cannizzaro intermediate (3), since it is a hydride ion donor, could also react with water to form HCO_2^- and H_2 (Equation 2).

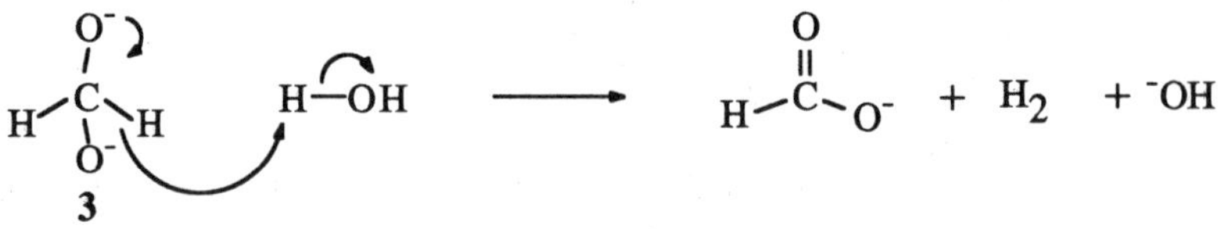

There is a report in the literature that very small amounts of H_2 are produced when a solution of formaldehyde in concentrated sodium hydroxide is heated *(7)*. Siemer (8) at Westinghouse Idaho Nuclear Co. demonstrated that significant amounts of H_2 (up to 26 % yield) could be produced by using lower concentrations of formaldehyde and higher concentrations of base. In Table II are shown the results of two experiments at low formaldehyde concentration and high concentrations of base.

Table II. Product Yields from the Reaction of Formaldehyde with
NaOH in Water[a]

T, °C	M, CH_2O	M, NaOH	%H_2	%HCOO⁻	%CH_3OH
28	4×10^{-4}	19	102	97	0
90	2×10^{-2}	17	41	71	27

[a]All reactions were carried out until hydrogen was no longer produced. Yields are based on formaldehyde.

The data show that, at room temperature and at 4×10^{-4} M formaldehyde, one molecule of formate was produced per molecule of hydrogen evolved and H_2 was produced in quantitative yield (Equation 2). When the concentration of formaldehyde was 2×10^{-2} M at 90° C, some methanol was produced (Equation 1). When the amount of methanol produced (27%) was subtracted from the total yield of formate (71%), the yield of formate produced by the hydrogen-producing reaction was obtained, and is approximately equal to the hydrogen yield (41%). In the second experiment, hydrogen formation and the Cannizzaro reaction are competing.

The kinetic order with respect to formaldehyde for the hydrogen formation reaction was determined at 28° C and 90° C from initial rate studies. The data are summarized in Table III. At both temperatures the kinetic order is approximately one, indicating that only one molecule of formaldehyde is participating in the rate-determining step of the reaction in which hydrogen is formed.

Table III. Order in Formaldehyde for the Reaction with NaOH
at Different Temperatures[a]

M, NaOH	M, CH_2O	T, °C	order[b]	k, $M^{-1} h^{-1}$
18	0.02 to 1	28 ± 1	1.1	$(7.0 \pm 0.3) \times 10^{-5}$
7	0.02 to 0.1	90 ± 1	0.9	$(3.0 \pm 0.2) \times 10^{-2}$

[a]All reactions were carried out to 5-10% conversion. [b]These values were calculated from the initial rates of H_2 evolution at different concentrations of CH_2O, taking the slope of the line from the plot ln(initial rate) vs. ln($[CH_2O]$). H_2 was measured by G.C., using a molecular sieve 5Å (60/80 mesh) column at r.t.

The kinetic order with respect to OH⁻ was similarly determined at 90° C from initial rate studies. The data are given in Table IV.

Table IV. Order in Hydroxide and Effect of Base Concentration on the Hydrogen Yield

M, CH$_2$O	M, NaOH	% H$_2$[a]	order[b]
1.5 x 10^{-3}	4	31	
1.5 x 10^{-3}	8	57	1.0
1.5 x 10^{-3}	16	98	

[a]All reactions were carried out at 60° C until hydrogen was no longer produced. Yields based on formaldehyde. [b]This value was calculated from the initial rates of H$_2$ evolution at different concentrations of NaOH, taking the slope of the line from the plot ln(initial rate) vs. ln(activity OH⁻). H$_2$ was measured as stated in Table III.

The order obtained from the slope of a plot, ln(initial rate) vs. ln(activity OH⁻) (9), was found to be 1.0. It is known that the equilibrium between CH$_2$O and CH$_2$(OH)$_2$ is strongly in favor of the gem-diol (**1**) (10). The equilibrium constant relating **1** and **2** in eqn. (2) is reported to be 6.0 at 50° C, (11) and can be calculated to be 5.4 at 25° C from the reported K$_a$ for the diol. (12) Thus, in these experiments, formaldehyde is present predominantly (>96%) as the CH$_2$(OH)O⁻ ion. It may be concluded from the first order dependence on [OH⁻] that CH$_2$O$_2^{2-}$ is the form of formaldehyde that acts as the hydride donor.

The rates of hydrogen generation for CH$_2$O and CD$_2$O were investigated at room temperature. The initial rate with protio formaldehyde was approximately 5 times faster than with deuterated formaldehyde (Figure 1). This isotopic effect suggests that the rate-determining step at room temperature involves cleavage of the C-H bond. Since OD⁻/D$_2$O is a stronger base system than OH⁻/H$_2$O, the concentration of Cannizzaro intermediate should be greater in the former system compared to the latter, and the observed rate differences between the deutero and protio compounds should represent a minimum.

The effect of the concentration of formaldehyde on hydrogen yield was studied at room temperature and at 60° C. The results are summarized in Table V. The hydrogen yield increased as the formaldehyde concentration decreased, which is consistent with the hydrogen-producing reaction (Equation 2) being unimolecular with respect to formaldehyde and thus competing more favorably at low concentrations of CH$_2$O with the Cannizzaro (bimolecular) reaction.

It was determined that as the concentration of base increases, the hydrogen yield increases (Table IV). These results are attributed to the fact that when [OH⁻] increases, the concentration of free aldehyde (which is necessary for the Cannizzaro reaction) decreases (Scheme 1), thus allowing the hydrogen-producing reaction (Equation 2) to compete even at high formaldehyde concentration.

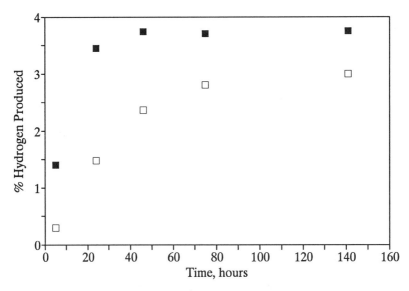

Figure 1. Hydrogen Evolution: CH_2O vs. CD_2O (15 M NaOH, 0.05 M CH_2O, T = 25° C; ■ CH_2O, □ CD_2O).

Table V. Yield of Hydrogen at Different Concentrations of CH_2O[a]

M, CH_2O	2×10^{-1}	2×10^{-2}	2×10^{-3}	4×10^{-4}	4×10^{-5}
%H_2[b]	1.4	8.2	34	61	104
%H_2[c]	6.5	31	76	102	99

[a]All reactions were carried out until hydrogen was no longer produced. Yields are based on formaldehyde. [b]T = 60°C, 11 M NaOH. [c]T = r.t., 19 M NaOH.

The proposed mechanism for the formation of H_2, based on these results, involves the formation of the Cannizzaro intermediate **3**, which reacts as a hydride ion donor with a water molecule to produce hydrogen, sodium formate and OH⁻ (Equation 2). The present data clearly indicate that when CH_2O and NaOH are allowed to react under conditions normally used in preparative reactions, the Cannizzaro products, HCOO⁻ and CH_3OH, are formed. Under these conditions only small amounts of H_2 are produced as was reported earlier *(7)*. However, when the concentration of base is high and the concentration of formaldehyde is low, the probability of a bimolecular reaction of the Cannizzaro intermediate with a formaldehyde molecule is small. Instead, intermediate **3** reacts with water to produce H_2. In these experiments concentrations of NaOH higher than the 2 M typical of the simulated waste have been used to enhance the hydrogen formation reaction, relative to the Cannizzaro reaction, so that the factors affecting the former reaction could be more easily established.

One of the ways of establishing whether or not the Cannizzaro intermediate **3** will react with water to produce hydrogen is to allow formaldehyde (CH_2O) to react with NaOD in D_2O and determine if HD is formed. However, due to exchange *(13)* of HD in D_2O/OD^- to produce D_2, this method did not provide reproducible data at the high base concentrations used *(30)*.

It is interesting to note that other aldehydes that do not possess alpha hydrogen atoms produce a significant amount of hydrogen in the presence of base (Table VI). In each of these cases the yield of hydrogen results from a competition between the Cannizzaro reaction and the H_2 formation reaction, both reactions involving the same hydride donor intermediate. It is interesting to note that although propanal has two alpha hydrogens, 1.6% H_2 was still produced.

Table VI. Hydrogen Yields from Aldehydes in the Presence of Base[a]

Aldehyde	% H_2
Formaldehyde	41, 43[b]
Glyoxylate ion	29
Pivaldehyde	30
Benzaldehyde	17
Propanal	1.6

[a]$T = 90°$ C. 17 M NaOH, 0.017 M RCHO in all cases. All reactions were carried out until hydrogen was no longer produced. Yields are based on the amount of the corresponding starting aldehyde. [b]Duplicate runs.

Finally, since we have observed a small decrease in the reaction rate when the radical inhibitor p-hydroquinone is added, a radical component to the overall mechanism must be considered. Therefore, the transfer of hydrogen from the Cannizzaro intermediate **3** to H_2O might take place as a hydrogen atom (H·) transfer (Equation 3), as well as a hydride ion (H^-) transfer.

$$H-\overset{O^-}{\underset{O^-}{\overset{|}{\underset{|}{C}}}}-H + H-OH \xrightarrow[\text{transfer}]{\text{electron}} H-\overset{O\cdot}{\underset{O^-}{\overset{|}{\underset{|}{C}}}}-H \quad H· \quad OH^- \longrightarrow H-\overset{O}{\overset{\|}{C}}-O^- + H_2 + OH^- \quad (3)$$

      3   **4**

Evidence for the formation of either of the radicals **4** or **5** and conversion to H_2 is found in the reactions of H_2O_2 and of Cu(II) with formaldehyde in base.

$$\begin{array}{cc} \text{O}^- & \text{OH} \\ | & | \\ \text{HCH} & \text{HCH} \\ | & | \\ \text{O}\cdot & \text{O}\cdot \\ \textbf{4} & \textbf{5} \end{array}$$

When formaldehyde in base reacts with H_2O_2, hydrogen and formate are produced rapidly and in quantitative yield (Equation 4) *(14)*.

$$2\ CH_2O + H_2O_2 + 2\ NaOH \rightarrow 2\ NaHCO_2 + H_2 + 2\ H_2O \qquad (4)$$

The mechanism for this reaction (Scheme 2) was first suggested by Wieland and Winglers *(15)*, and has been supported by the work of Wirtz and Bonhoeffer *(16)*, Jaillet and Quellet *(17)*, and Abel *(18)*. The intermediate **6** has been prepared independently *(15,17)* and shown to react in base to give formate and hydrogen in a 2:1 ratio. In this case both hydrogen atoms in H_2 come from formaldehyde.

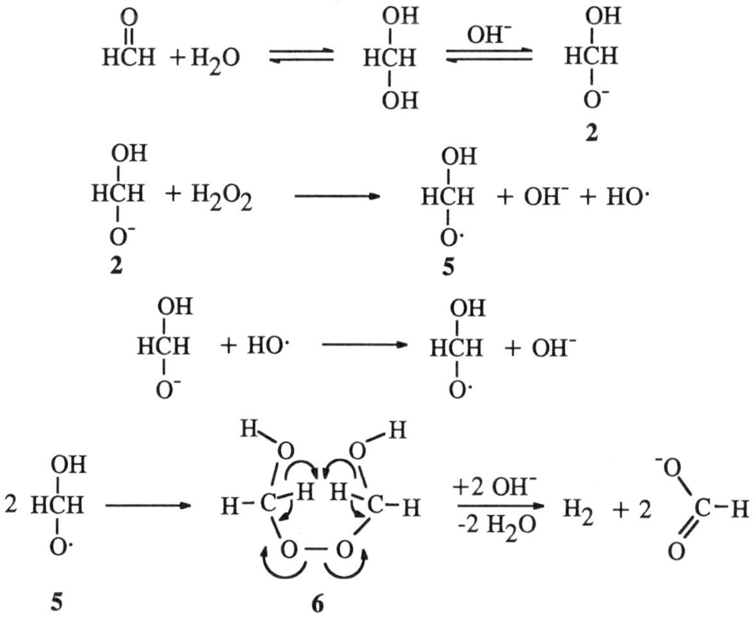

Scheme 2. Proposed Mechanism for the Reaction of Formaldehyde with NaOH in the Presence of Hydrogen Peroxide.

The rapid formation of hydrogen from formaldehyde by reaction with Cu(II) ion (Figure 2) can be interpreted in terms of a similar mechanism. These data were obtained using a Teflon vessel connected to a buret with a mercury seal which allowed measurement of the amount of gas produced. Into the reaction vessel was placed 300 mL of 20 M sodium hydroxide which was heated at 90° C with stirring. While stirring, 1.4 mL of a solution containing 1.63 mmoles of formaldehyde and 3.0×10^{-2} mmoles of $CuCl_2$ were added through a septum by means of a syringe provided with a stainless steel needle. The characteristic brown color due to Cu(0) was immediately observed. The difference between the amounts of hydrogen produced in the presence and absence of Cu(II) is approximately 7 mL (~0.23 mmoles) throughout the experiment, indicating an initial rapid formation of hydrogen.

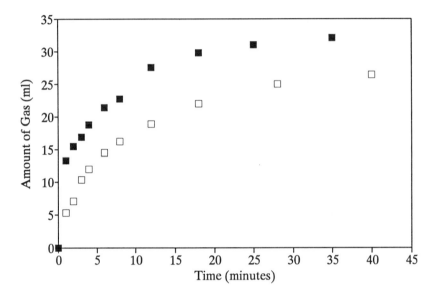

Figure 2. Reaction of Formaldehyde with NaOH. Effect of Cu(II) on Gas Evolution (■ $CH_2O/Cu(II)$, □ CH_2O).

A mechanism suggested to explain the initial rapid formation of H_2 (Scheme 3) involves the formation of the Cannizzaro intermediate **3** which then reduces Cu(II) to Cu(0) and forms the intermediate radical **4** which can dimerize to **6**. Dimer **6** can then collapse to form H_2 and formate *(7)*. In this case, as with hydrogen peroxide, both hydrogen atoms originate from formaldehyde. This point however, has not been established. Copper metal can, depending on its physical state, serve as a heterogeneous catalyst for the formation of hydrogen. *(7,19)* Some contribution to hydrogen formation in the initial period can be attributed to this effect. Mechanisms involving electron transfer are then suggested for three reactions (with H_2O_2, Cu(II), and H_2O) of the Cannizzaro intermediate leading to the formation of formate and hydrogen. The mechanism of hydrogen formation thus, can be polar in nature, radical in nature, or both in competition.

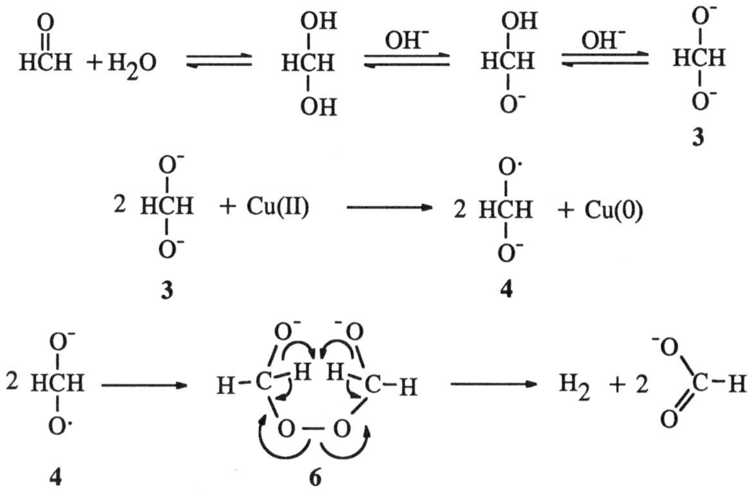

Scheme 3. Proposed Mechanism for the Reaction of Formaldehyde with NaOH in the Presence of Cu(II).

Regardless of the mechanistic details, the formation of hydrogen by reaction of formaldehyde, or other aldehydes, in concentrated NaOH seems well established, as well as the factors favoring the occurrence of that reaction. Since it is reasonable to assume that formaldehyde is present in some small, steady-state concentration in tank 101-SY, the reaction may make a significant contribution to hydrogen being formed.

Source of the Nitrogen in the N_2O and N_2 Formed in the Thermal Decomposition of HEDTA *(27)*

There are other gases in addition to H_2 that are produced in significant amounts when HEDTA is heated to $120°$ C in a simulated waste mixture. These gases are N_2O, N_2, NH_3, CH_4 and CO_2 (CO_2 reacts with NaOH immediately to form Na_2CO_3), and they form as a result of a series of reactions. The question concerning the formation of N_2O, N_2 and NH_3 is, "do the nitrogen in these gases have their sources in NO_2^- and NO_3^- or is the source of the nitrogen in HEDTA"? This question was answered by isotopically labeling the nitrogen atom in NO_2^- and NO_3^- as ^{15}N and determining whether or not the N_2O and N_2 formed contain the ^{15}N label. If the N_2O and N_2 formed are doubly-labeled with the ^{15}N label, then these gases must have originated from the [15]NO_2^- and/or [15]NO_3^-; on the other hand, if the N_2O and N_2 formed did not contain the ^{15}N label, then the N_2O and N_2 must have come from the unlabeled HEDTA.

Several reactions of unlabeled HEDTA in SY1-SIM-91B media prepared with ^{15}N-labeled nitrate and/or nitrite reagents were conducted at 120° C for 49 days under atmospheres of air and helium using the apparatus shown in Figure 3. Mass spectrometric analyses of the head space gases clearly established that the N_2O and N_2 produced in these reactions originated only from the nitrite and not from nitrate or HEDTA. For example, when the simulated waste mixture contained both [15]NO_2^- and [15]NO_3^-, and the reaction was conducted under either an air or a helium atmosphere, only doubly labeled N_2O and N_2 were formed.

On the other hand, when HEDTA was allowed to react at 120° C in the simulated waste mixture containing [15]NO_3^- as the only ^{15}N label, only unlabeled N_2O and N_2 were formed. This result indicates that both N_2O and N_2 originate from NO_2^-. In order to verify this conclusion, HEDTA was allowed to react at 120° C in the simulated waste mixture containing [15]NO_2^- as the only nitrogen labeled compound. In this case both doubly-labeled [15]N_2O and [15]N_2 were produced. This result shows conclusively that both N_2O and N_2 originate exclusively from NO_2^-.

It would be difficult to prepare ^{15}N-HEDTA in order to check the above results that show that N_2O and N_2 do not originate from the organic portion of the simulated waste; however, it is not necessary to do this since, if any N_2O and N_2 did originate from HEDTA, then some [14]N - N_2O and /or N_2 would have been produced in the reaction with [15]NO_2^-, which it was not.

There are, of course, many organic degradation products produced in the thermal decomposition of HEDTA in simulated waste at 120° C; glycine is one of these. Since ^{15}N-glycine was available, the decomposition of this compound was studied in simulated waste at 120° C and a trace of [15]N_2 in the ^{14}N-N_2O, N_2 mixture was found. Great significance is not attached to detection of a trace amount of [15]N_2 in this experiment and therefore NO_2^- is still considered the major source of N_2O and N_2 produced in simulated waste. All these reactions were carried out in the apparatus shown in Figure 3.

Source of Nitrogen in NH_3 Formed in the Thermal Decomposition of HEDTA (27)

Formation of ammonia in these thermal degradation studies arises after a series of reactions. When HEDTA was allowed to react at 120° C for 49 days in simulated waste, NH_3 was formed in 17% yield (based on HEDTA). The NH_3 was passed into an aqueous solution of HCl and the ammonia determined as NH_4Cl. When HEDTA was allowed to react at 120° C in simulated waste containing [15]NO_2^- and [15]NO_3^-, both [15]NH_3 (90%) and [14]NH_3 (10%) were formed, as indicated from ^{14}N and ^{15}N NMR spectra of the NH_4Cl solution. When a similar reaction was carried out containing only [15]NO_3^- as the labeled compound, only unlabeled NH_3 was formed; however, when only [15]NO_2^- was present, [15]NH_3 (88%) and [14]NH_3 (12%) were formed. All of these results clearly indicate that approximately 90% of the NH_3 produced in the thermal decomposition of HEDTA in simulated waste at 120° C over a period of 49 days originates from $NaNO_2$ and 10% originates from HEDTA and its decomposition products. All these reactions were carried out in the apparatus shown in Figure 3.

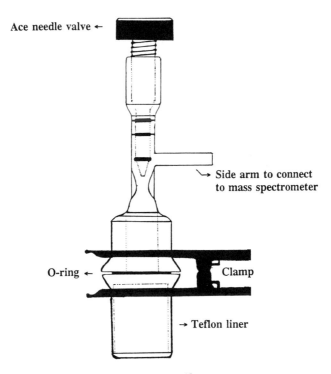

Figure 3. Apparatus for ^{15}N experiments.

Kinetic and Mechanistic Studies (26)

It is very important for the purpose of understanding the chemistry in 101-SY, and for remediation, that the rate at which gases are evolved and the mechanisms whereby these gases are formed, be well understood. Therefore, an investigation of the kinetics, mechanism and reaction products of the thermal degradation of certain chelating agents (HEDTA, glycolate) in an aqueous solution containing $NaNO_2$, $NaNO_3$, NaOH, $NaAlO_2$ and Na_2CO_3 was conducted. An analysis of the rates of formation of the volatile and nonvolatile thermal degradation products was conducted and the rate of decomposition of each of the reactants was studied. The concentrations of reactants employed in all reported studies are given in Table I. The reactions components were added to a 125 mL Erlenmeyer flask containing approximately 50 mL of distilled H_2O. The mixture was stirred until the solids dissolved. The solution was then filtered through a fritted glass funnel to eliminate any residual suspended matter and subsequently transferred quantitatively to a 100 mL volumetric flask. The solution was brought to the mark with distilled H_2O, allowed to stand at room temperature overnight, and was then transferred into the reaction vessel. The vessel was heated to 120 °C and gas volume measurements were taken. The reaction was allowed to continue for periods varying from 300 to 1700 h.

It should be emphasized that these reactions were carried out under an air atmosphere. In addition, production of gas often occurred after an induction period of variable length depending on the reaction components. The rates of gas evolution were calculated from the slope of the best straight line in the plots of mL of gas vs. time. The apparatus used to make the volumetric determinations of gas evolved is given in Figure 4.

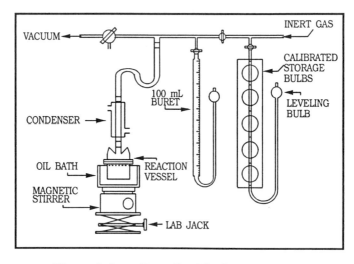

Figure 4. Long Term Gas Kinetics Apparatus.

Sodium Glycolate as the Organic Component *(28)*

A series of experiments was conducted utilizing glycolate as the organic component in which each of the inorganic reagents was sequentially omitted in order to determine the effect of the particular inorganic reagent on gas evolution. The results of these experiments are summarized in Table VII.

The rate of gas evolution with all components present under an air atmosphere or a helium atmosphere (utilizing deoxygenated water) was the same (~ 0.34 mL/hour). When glycolate was omitted from the reaction mixture, there was no appreciable gas evolution. When $NaNO_2$ was omitted from the reaction mixture, there was an insignificant amount of gas evolved.

Table VII. Average Rate of Gas Evolution With Glycolate as the Organic Component With an Altered Set of Standard Conditions

Component Omitted From Standard Mixture	Induction Period (hours)	Rate (mL/h)
none (He atmosphere)	20	0.34
none (air atmosphere)	50	0.34
glycolate	0[a]	0
$NaNO_2$	155	$\approx 0.05^c$
$NaNO_3$	72	0.3
$NaAlO_2$	0[b]	0
Na_2CO_3	150	0.28

[a]Experiment run for 437.5 h. [b]Experiment run for 214 h. $^c \approx$ 20-mL of gas was evolved over 400 h.

When $NaNO_3$ or Na_2CO_3 was omitted from the reaction mixture, gas evolution was approximately the same as in the case when it was present. When $NaAlO_2$ was omitted from the reaction mixture, there was no appreciable gas evolution. This investigation showed the importance of $NaAlO_2$ and $NaNO_2$ in the formation of gas, relative to the other inorganic components in the mixture. Analysis of the nonvolatile thermal degradation products from these reactions showed the disappearance of glycolate and nitrite and the formation of formate and oxalate. A quantitative investigation of the nonvolatile thermal degradation components of glycolate in the standard mixture over time was conducted using ion chromatography (IC). The results from the analysis for nitrite, nitrate, oxalate, formate and glycolate are summarized in Table VIII.

Table VIII. Nonvolatile Reaction Components of Glycolate Thermal Degradation

Time (h)	Nitrite (mol x 10^2)	Nitrate (mol x 10^2)	Oxalate (mol x 10^4)	Formate (mol x 10^4)	Glycolate (mol x 10^3)
0	5.60	6.48	0	0	5.41
77	5.50	6.63	<0.5	1.22	4.78
197	5.38	6.58	5.89	4.03	3.95
310	5.35	6.30	4.93	5.22	3.37
406	5.33	6.55	8.76	5.99	2.93
514	5.30	6.50	10.5	6.77	2.66
672	5.30	6.37	15.2	7.16	1.96
981	5.05	6.37	22.1	8.25	0.459
1250	5.03	6.47	26.5	8.85	0.243
1474	5.00	6.33	27.7	9.23	0.269
1582	4.92	6.65	31.1	9.26	0.111

The number of moles of nitrate does not appear to change (within experimental error) during the course of the experiment. The level of nitrite decreased by 6.8 x 10^{-3} moles ($\approx$ 12%). Oxalate (3.11 x 10^{-3} moles) and formate (9.26 x 10^{-4}) were formed over this same time period. The level of glycolate decreased by 5.30 x 10^{-3} moles ($\approx$ 98%). A first order plot for the disappearance of glycolate (ln[glycolate] vs time) is shown in Figure 5. From this plot, a rate constant k = 1.5 x 10^{-3} h^{-1} and a first order half life of 462 hours were calculated. This value was independently confirmed by monitoring the disappearance of ^{13}C labeled glycolate (at both carbons) by ^{13}C NMR. The proposed mechanism for the decomposition of glycolate in simulated waste is as follows (Equations 5-16):

$$Al(OH)_4^- + NO_2^- \rightleftharpoons Al(OH)_3\text{-}O\text{-}NO^- + OH^- \tag{5}$$

$$Al(OH)_3\text{-}O\text{-}NO^- + HO\text{-}CH_2\text{-}CO_2^- \rightleftharpoons Al(OH)_4^- + ON\text{-}O\text{-}CH_2\text{-}CO_2^- \tag{6}$$

$$ON\text{-}O\text{-}CH_2\text{-}CO_2^- \rightarrow NO^- + CH_2O + CO_2 \tag{7}$$

$$ON\text{-}O\text{-}CH_2\text{-}CO_2^- + OH^- \rightarrow NO^- + H\text{-}(CO)\text{-}CO_2^- + H_2O \tag{8}$$

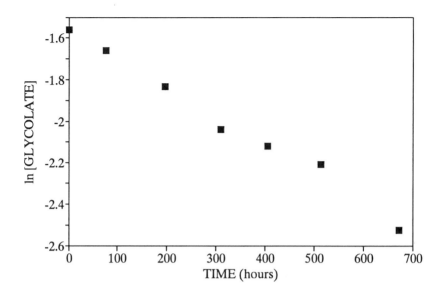

Figure 5. First Order Plot for the Disappearance of Glycolate.

$$NO^- + NO_2^- \rightleftharpoons N_2O_3^{2-} \tag{9}$$

$$NO^- + NO^- \rightarrow N_2O_2^{2-} \tag{10}$$

$$N_2O_2^{2-} + H_2O \rightleftharpoons HN_2O_2^- + OH^- \tag{11}$$

$$HN_2O_2^- \rightarrow N_2O + OH^- \tag{12}$$

$$CH_2O + OH^- \rightleftharpoons HO\text{-}CH_2\text{-}O^- \tag{13}$$

$$^-O\text{-}CH_2\text{-}O^- + H_2O \rightarrow H_2 + H\text{-}COO^- + OH^- \tag{14}$$

$$H\text{-}(CO)\text{-}CO_2^- + OH^- \rightleftharpoons {}^-O\text{-}CH(OH)\text{-}CO_2^- \rightleftharpoons (^-O)_2\text{-}CH\text{-}CO_2^- \tag{15}$$

$$(^-O)_2\text{-}CH\text{-}CO_2^- + H_2O \rightarrow H_2 + {}^-O_2C\text{-}CO_2^- + OH^- \tag{16}$$

If it is assumed that the decomposition of the nitrite ester of glycolate (Equation 7) is fast compared to its rate of formation (Equation 6), then all steps following the decomposition can be assumed to be fast and therefore can be excluded from the kinetic analysis. Based on these assumptions a rate (Equation 17) can be derived:

$$Rate = \frac{k[Gly]^n[AlO_2]^m[NO_2]^p}{[OH^-]^q} \tag{17}$$

where $n=1$ and m, p and q are to be determined. Since glycolate is present in small quantities compared to aluminate, nitrite and hydroxide, the disappearance of glycolate should follow pseudo first-order kinetics (Equation 18).

$$ln[Gly]_t - ln[Gly]_{t=0} = k't \tag{18}$$

Thus, a plot of ln [Glycolate] vs. time should give a straight line with slope equal to k', (Equation 19),

$$k' = \frac{k[AlO_2^-]^m[NO_2^-]^p}{[OH^-]^q} \tag{19}$$

where this derivation assumes m and p order dependence in AlO_2^- and NO_2^-, respectively. Rearranging the above expression, (Equation 20) produces:

$$\frac{k}{[OH^-]^q} = \frac{k'}{[AlO_2^-]^m[NO_2^-]^p} \tag{20}$$

A series of experiments was conducted under an air atmosphere in order to ascertain the effect of changes in the concentration of AlO_2^-, glycolate and NO_2^- in simulated waste on $k/[OH^-]^q$. In the first series of experiments, the concentration of aluminate was reduced by one-third. In the second series of experiments, the concentration of glycolate was halved and in the third series, the concentration of nitrite was halved. The reaction solutions were pipetted into Teflon lined glass reaction vessels and heated at 120° C. Reaction vessels were removed from the heat after 100, 200 and 300 hours and their contents were analyzed by ion chromatography. It must be stated that although the reactions were conducted in Teflon lined glass vessels, there was still some contact with the glass. If it is assumed that the reaction is first-order with respect to AlO_2^- and NO_2^-, then the $k/[OH^-]^q$ values for each series should be identical. From the data in Table IX it can be seen that the $k/[OH^-]^q$ values are very nearly the same, from which it can be concluded that the reaction is first order in glycolate, aluminate and nitrite.

Key steps in the proposed reaction scheme involve the generation of an aluminum complexed nitrite species and its subsequent reaction with glycolate to generate a nitrite ester. The nitrite ester subsequently undergoes a fragmentation process (Equation 7) to produce CO_2, NO^- and formaldehyde or a β-elimination (Equation 8) to produce glyoxylate and NO^-. We have already shown that formaldehyde and glyoxylate are key intermediates in the formation of formate, oxalate and H_2 (equations 13-16). Equations 9-12 are well known (20). In order to show that the nitrite ester does indeed give the postulated fragmentation products suggested in the mechanism, the nitrite ester of methyl glycolate was synthesized (Equation 21) based on a procedure by Doyle (21). A mixture of methyl glycolate and the nitrite ester of methyl glycolate was isolated in a $\approx$ 4:1 ratio. The isolated mixture of product and starting material was then introduced into a simulated waste solution. The nitrite ester was found to decompose to H_2, N_2O, CO_2, NO_2^-, formate and oxalate.

Table IX. Rate Constants for Thermal Degradation of Glycolate under Air

Experiment	k' (h^{-1})	$k/[OH^-]^q$ ($h^{-1}M^{-2}$)
0.21 M Glycolate 1.54 M Aluminate 2.24 M Nitrite 0.42 M Carbonate 2.59 M Nitrate 2.00 M Hydroxide	$(6.5 \pm 1.3) \times 10^{-4}$	$(1.9 \pm .4) \times 10^{-4}$
0.21 M Glycolate 1.00 M Aluminate 2.24 M Nitrite 0.42 M Carbonate 2.59 M Nitrate 2.00 M Hydroxide	$(4.5 \pm 1.0) \times 10^{-4}$	$(2.0 \pm .4) \times 10^{-4}$
0.10 M Glycolate 1.54 M Aluminate 2.24 M Nitrite 0.42 M Carbonate 2.59 M Nitrate 2.00 M Hydroxide	$(7.0 \pm 1.5) \times 10^{-4}$	$(2.0 \pm .4) \times 10^{-4}$
0.21 M Glycolate 1.54 M Aluminate 1.12 M Nitrite 0.42 M Carbonate 2.59 M Nitrate 2.00 M Hydroxide	$(3.7 \pm 1.0) \times 10^{-4}$	$(2.1 \pm .4) \times 10^{-4}$

$$t\text{-BuONO} + HOCH_2COOCH_3 \rightleftharpoons t\text{-BuOH} + ONOCH_2COOCH_3 \quad (21)$$

All of the kinetic work with glycolate was carried out in glass vessels and is now being repeated in Teflon vessels.

HEDTA as the Organic Component *(29)*

The preceding work has provided the basis for further investigations into more complex systems. The mechanism which was proposed for glycolate decomposition can also be applied to the thermal degradation of HEDTA. A preliminary study of the composition of the nonvolatile degradation products over time has been completed (Table X). This initial work should provide a foundation for a thorough analysis of the nonvolatile thermal degradation products.

Table X. Nonvolatile Reaction Components of HEDTA from
Thermal Degradation

Time (h)	Nitrite (mol x 10^2)	Nitrate (mol x 10^2)	Formate (mol x 10^3)	HEDTA (mol x 10^3)
0	5.60	6.65	0	5.30
112.5	5.48	6.55	0.48	5.08
191	5.43	6.57	1.01	4.84
303	5.40	6.50	1.02	4.47
408.5	5.38	6.74	1.30	4.03
500.5	5.33	6.64	1.89	3.60

The trends observed in the thermal degradation of HEDTA are the same as those observed in the thermal degradation of glycolate. The level of nitrate does not change (within experimental error) during the course of the experiment. The level of nitrite decreases by 2.7 x 10^{-3} moles ($\approx$ 4.8%). Formate (1.89 x 10^{-3}) is formed over this same time period. The level of HEDTA decreases by 1.7 x 10^{-3} moles ($\approx$ 32%).

Figure 6 shows a plot of ln [HEDTA] vs time. Unlike the result with glycolate, HEDTA does not disappear in a first order fashion, which is not surprising since HEDTA can react in several different ways.

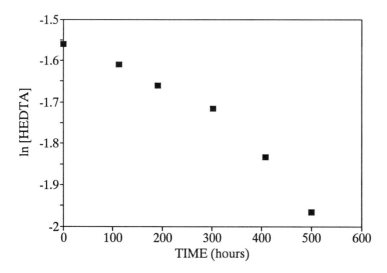

Figure 6. Long Term HEDTA Degradation Study. Ln[HEDTA] vs. Time in Simulated Waste.

Identification of HEDTA Degradation Products using ^{13}C NMR *(29)*

Another objective of the Georgia Tech effort has been to utilize the earlier studies done at Hanford and Pacific Northwest Laboratories as the basis for further development of an understanding of both the stoichiometry and the temporal formation of individual degradation products of organic complexants in waste tank 101-SY. The ^{13}C studies of doubly ^{13}C-labeled glycolic acid and of two forms of labeled HEDTA (HEDTA-A and HEDTA-B; asterisks indicate ^{13}C-labeled carbons) described below have been especially useful in this regard. The use of compounds that are highly enriched with ^{13}C in specific positions allows for the detection of products early in the reaction and the spin-spin coupling patterns arising from 1,2-labeled derivatives are very useful in assigning the products.

HEDTA-A   **HEDTA-B**

HEDTA-A was prepared by the following reaction sequence:

$$^{13}CH_2{}^{13}CH_2 + H_2NCH_2CH_2NH_2 \rightarrow HO^{13}CH_2{}^{13}CH_2NHCH_2CH_2NH_2 \qquad (22)$$
$$\diagdown O \diagup$$

$$HO^{13}CH_2{}^{13}CH_2NHCH_2CH_2NH_2 + 3\ Na[BrCH_2CO_2] \rightarrow HEDTA\text{-}A \qquad (23)$$

The product was purified by fractional crystallization from methanol. HEDTA-B was prepared by exhaustive alkylation of unlabeled $HOCH_2CH_2NHCH_2CH_2NH_2$ with $Na[Br^{13}CH_2{}^{13}CO_2]$, i.e.,

$$HOCH_2CH_2NHCH_2CH_2NH_2 + 3\ Na[Br^{13}CH_2{}^{13}CO_2] \rightarrow HEDTA\text{-}B \qquad (24)$$

To identify the products of complexant degradation it was necessary to establish a data base of ^{13}C spectra of authentic samples of HEDTA, EDTA, glycolate, and known, or probable, degradation products. Commercial samples were used when available, others were synthesized as follows. U-EDDA *(22)* and EDMA *(23)* were synthesized using reported procedures. N-Methyl-S-EDDA was prepared by the methylation of S-EDDA lactam followed by base hydrolysis. N-Methyl-ED3A was synthesized by exhaustive alkylation of N-methylethylenediamine using sodium bromoacetate. N-Methyl-U-EDDA was isolated from the reaction of N-methylethylenediamine and sodium chloroacetate. EAMA was synthesized from the reaction of ethanolamine and sodium bromoacetate. Carbon-13 NMR spectra of the chelating agents and possible fragments of thermal degradation in simulated waste solution were obtained and the chemical shifts were measured with respect to either

C_6D_6 in an external cell or sodium carbonate of the simulated waste solution. N-Methyl-U-EDDA reacted with simulated waste solution even at room temperature, so that the ^{13}C NMR spectrum was measured on a freshly prepared 2M NaOH solution. Spectral data for all compounds are given in Table XI.

The thermal reaction of ^{13}C-labeled HEDTA-A in simulated waste solution was carried out at 120° C for a total of 1452 hours in a Teflon container, capped with a Teflon disc, that was housed in a brass metal tube with a brass screw cap (Figure 7). One mL of solution was used in each experiment. The reaction mixture was transferred at regular intervals to a NMR tube using a pipette, and ^{13}C NMR spectra were recorded. Then, the solution was returned to the Teflon liner and heating was continued. Intensities of all peaks were measured relative to C_6D_6, which was in an external cell. The solution was diluted to 1 mL (starting amount) after 60.5 days of heating at 120° C and the NMR measured again to estimate the amount of remaining HEDTA (50%).

After 1.5 days the ^{13}C spectrum of the HEDTA-A reaction mixture contained a new resonance for formate and a more intense resonance for carbonate, which indicates formation of CO_2. Oxalate was detectable after 18.5 days and labeled glycine after 34.5 days. The spectrum recorded after 60.5 days also contained a singlet due to unlabeled glycine, indicating that its formation occurs from the carboxymethyl groups as well as from the hydroxyethyl group. The relative amounts of glycine formed from $-N^{13}CH_2{}^{13}CH_2OH$ and unlabeled $-NCH_2CO_2H$ after 60.5 days was approximately 1:6. Two pairs of doublets centered at 50.51 and 60.28 ppm (J=38 Hz) and 43.03 and 63.33 ppm (J=38 Hz) were detected after 13.5 days and 34.5 days, respectively, and have been assigned as ethanolamine-N-monoacetic acid (EAMA) and ethanolamine (EA), respectively, based on a comparison of their chemical shifts with those of authentic material. A summary of the appearance of products versus time data is given below and the ^{13}C NMR spectrum, obtained after 60.5 days of reaction, is given in Figure 8.

After   36h HCO_2^-, CO_2
        60h HCO_2^-, CO_2
     108h HCO_2^-, CO_2
     156h HCO_2^-, CO_2
     204h HCO_2^-, CO_2, EAMA (traces)
     444h HCO_2^-, CO_2, EAMA, $C_2O_4^=$
     564h HCO_2^-, CO_2, EAMA, $C_2O_4^=$
     828h HCO_2^-, CO_2, EAMA, $C_2O_4^=$, EA, Glycine (labeled)
   1452h HCO_2^-, CO_2, EAMA, $C_2O_4^=$, EA, Glycine (labeled and unlabeled)

The ^{13}C NMR studies involving the thermal decomposition of HEDTA-B in simulated waste at 120°C were carried out in a similar fashion to that described for HEDTA-A. Resonances in the ^{13}C NMR spectrum obtained after 1808 h (Figure 9) have been assigned to ED3A, U-EDDA, S-EDDA, IDA, glycine, formate, oxalate and carbonate. The assignment of resonances to ED3A is not straightforward as they overlap with resonances due to unreacted U-EDDA and S-EDDA. However, the spectrum also contained five doublets of low intensity at 50.86, 55.67, 169.93, 176.62

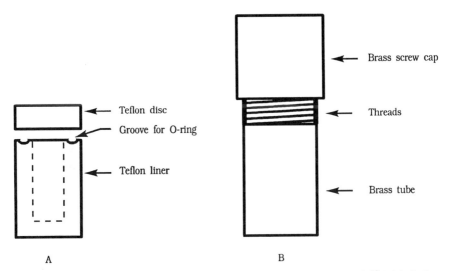

Figure 7. Reaction vessel used for thermal decomposition of ^{13}C-labeled compounds.

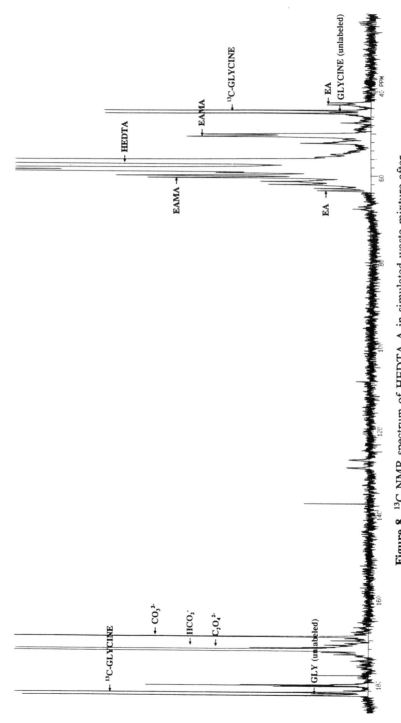

Figure 8. ^{13}C NMR spectrum of HEDTA-A in simulated waste mixture after heating at 120° C for 1452 h.

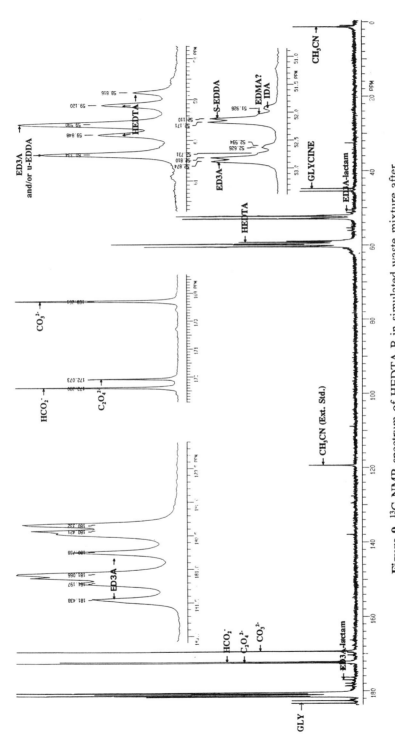

Figure 9. ^{13}C NMR spectrum of HEDTA-B in simulated waste mixture after heating at 120° C for 1808 h.

Table XI. ^{13}C Chemical Shifts (ppm) in Simulated Waste Solution[a]

EDTA	51.95	59.13	180.81					
HEDTA	51.47	52.09	56.75	58.45	59.11	59.39	180.71	180.99
ED3A[b]	45.71	52.50	54.99	59.92	180.62	180.99		
NTA	59.37	180.48						
IDA	52.19	180.61						
S-EDDA[b]	47.63	52.43	180.62					
U-EDDA	37.81	58.20	59.92	180.60				
N-ME S-EDDA[b]	42.31	45.63	52.57	56.31	61.89	179.56	180.56	
N-ME ED3A[b]	42.60	52.65	54.43	59.88	62.15	180.53	180.80	
HIDA	57.49	58.89	59.89	180.64				
DMG[b]	44.73	63.31	179.48					
GLY[b]	45.05	182.68						
GLYCOLATE	61.86	183.07						
N-ME GLY[b]	34.83	54.29	180.65					
N-ME IDA[b]	42.86	61.79	180.23					
ED	43.49							
N-ME ED	34.90	39.93	53.03					
HED	40.28	50.79	50.96	60.23				
EDMA[b]	40.06	50.73	52.30	180.56				
N-ME U-EDDA[c]	34.99	40.18	44.34	53.03	175.50			
EA	43.25	63.29						
EAMA	50.68	52.41	60.36	180.74				

[a]Chemical shifts are measured relative to an external C_6D_6 standard unless otherwise indicated. Abbreviations used in this table that were not included on page 3 are: N-ME S-EDDA, $HO_2CCH_2N(CH_3)CH_2CH_2NCH_2CO_2H$; N-ME-ED3A, $HO_2CCH_2N(CH_3)CH_2$-$CH_2N(CH_2CO_2H)_2$; HIDA, $HOCH_2CH_2N(CH_2CO_2H)_2$; DMG, $(CH_3)_2NCH_2CO_2H$; N-ME GLY, $CH_3NHCH_2CO_2H$; N-ME IDA, $CH_3N(CH_2CO_2H)_2$; ED, $H_2NCH_2CH_2NH_2$; N-ME ED, $CH_3NHCH_2CH_2NH_2$; HED, $HOCH_2CH_2NHCH_2CH_2NH_2$; EDMA, H_2NCH_2-$CH_2NHCH_2CO_2H$; N-ME U-EDDA, $CH_3NHCH_2CH_2N(CH_2CO_2H)_2$. [b]Chemical shifts for these compounds were measured with respect to sodium carbonate in simulated waste solution and converted to the C_6D_6 external standard scale. [c]Spectrum recorded in 2 M NaOH due to the sensitivity of N-ME U-EDDA to simulated waste solution.

and 178.25 ppm, which are most likely due to ED3A lactam ($\delta = 50.86$, 55.66, 60.50, 169.99, 176.53 and 178.15, for authentic material). The doublet at 60.50 ppm, observed for the authentic lactam, could not be seen as it lies under more intense resonances of HEDTA and U-EDDA. ED3A has been reported to cyclize to give the lactam under acidic conditions and to revert back to ED3A in basic medium *(24)*. Because of the strongly basic nature of the simulated waste solution, ED3A, if present, should be largely in the uncyclized form; therefore the presence of minor amounts of the lactam indicates the presence of substantial amounts of the uncyclized material. We therefore believe that the intense doublets at 59.96 ppm and 52.46 ppm must be partially due to ED3A, as well as U-EDDA, and S-EDDA. Further support for the assignment of these resonances is provided by the lowest field carboxylate resonance, observed at 181.08 ppm (J=52.6 Hz) in the 1808 hr spectrum. Only two compounds, HEDTA ($\delta = 180.71$ and 180.99) and ED3A ($\delta = 180.62$ and 180.99) have carboxylate resonances with chemical shifts > 180.90 ppm. The larger coupling constant of HEDTA (55.2 Hz) suggests that this doublet is substantially due to ED3A. It is worth noting that the intensity of the doublet at 52.43 ppm that is assigned to S-EDDA increased relative to the doublet at 52.50 assigned to ED3A during the course of the reaction. A summary of the assignments is give below and the ^{13}C NMR is given in Figure 9.

After	14h	little reaction
	37h	little reaction
	135h	HCO_2^-, CO_2, $C_2O_4^=$, ED3A &/or U-EDDA, S-EDDA
	232h	HCO_2^-, CO_2, $C_2O_4^=$, ED3A &/or U-EDDA, S-EDDA
	399h	HCO_2^-, CO_2, $C_2O_4^=$, ED3A &/or U-EDDA, S-EDDA
	562h	HCO_2^-, CO_2, $C_2O_4^=$, ED3A &/or U-EDDA, S-EDDA
	732h	HCO_2^-, CO_2, $C_2O_4^=$, ED3A &/or U-EDDA, S-EDDA
		IDA, ^{13}C-Glycine, ED3A Lactam
	1232h	HCO_2^-, CO_2, $C_2O_4^=$, ED3A &/or U-EDDA, S-EDDA
		IDA, ^{13}C-Glycine, ED3A Lactam
	1808h	HCO_2^-, CO_2, $C_2O_4^=$, ED3A &/or U-EDDA, S-EDDA
		IDA, ^{13}C-Glycine, ED3A Lactam

Identification of Irradiated HEDTA Degradation Products Using ^{13}C NMR Spectroscopy.

Samples of HEDTA-A and HEDTA-B were irradiated in simulated waste solution (50-60 mg in 5 mL solution contained in steel tubes) at doses of 19.6 Mrad and 20.82 Mrad, respectively at Argonne National Laboratory (D. Meisel) and were returned to Georgia Tech for ^{13}C NMR spectral analysis. As reported previously, HEDTA reacts significantly upon irradiation, as indicated by the low intensity of the resonances for HEDTA in these samples (Figures 10 and 11). Tentative assignments of resonances are shown in the spectra of the irradiated samples.

Because time dependent irradiation studies are not available it is not possible to draw definitive conclusions concerning differences in products or their distributions

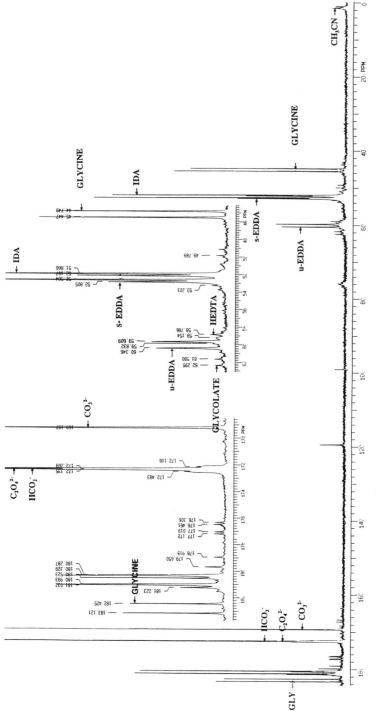

Figure 10. ^{13}C NMR spectrum of HEDTA-A after irradiation in simulated waste mixture (19.6 Mrad dose).

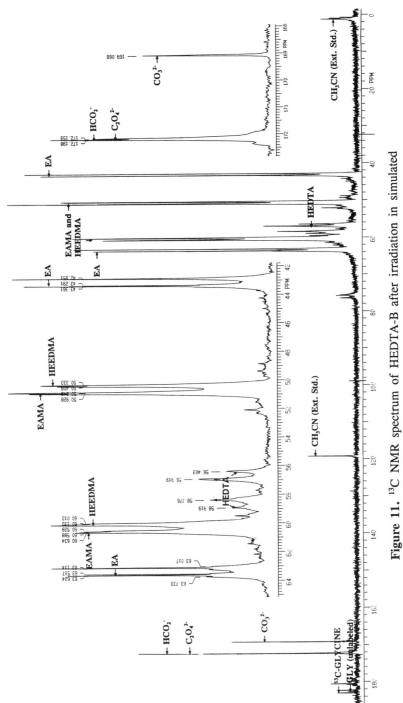

Figure 11. ^{13}C NMR spectrum of HEDTA-B after irradiation in simulated waste mixture (20.8 Mrad dose).

between the thermal and irradiative degradation pathways. One obvious difference however, is that a substantial amount IDA is formed as one of the products derived from irradiation of HEDTA-B, whereas IDA has not been observed as a significant product from thermal reactions. However, it must be noted that much more HEDTA has been consumed in the irradiation of HEDTA-B than in the corresponding thermal reaction. Large amounts of IDA were detected by Lokken, et al, in Tank-107AN;(2) thus it could be conjectured that radiolytic decomposition of HEDTA could be a major pathway for the generation of IDA in the waste storage tank.

Glycine is formed in substantial amounts in radiolytic degradation as it is from the thermal reaction. The ^{13}C spectrum of HEDTA-A contained a significant carboxylate resonance for unlabeled glycine which indicates that a substantial amount of glycine is produced from the carboxymethyl groups (Figure 11). This observation is confirmed by the detection of intense doublets at 45.09 and 182.77 ppm (J=52.6 Hz) in the spectrum of the carboxymethyl labeled compound that are due to glycine. The low intensity of the ^{13}C-glycine signals in the spectrum of the hydroxyethyl labeled compound indicates that only a small amount of glycine is produced from the hydroxyethyl part of the molecule. Similar observations were made in the thermal decomposition of these labeled materials.

Doublets are present in the spectrum of HEDTA-B at 52.45 ppm (J=52.7 Hz) and 59.97 ppm (J=55.6 Hz) that are most likely due to S-EDDA and U-EDDA, respectively. ED3A has resonances at comparable chemical shifts but the absence of a low field doublet at ca. 180.99 ppm or resonances for the lactam suggest that there is little of this compound present.

The spectrum of HEDTA-B also contains a low intensity doublet at 61.94 ppm, which may be due to glycolate. The intensity of resonances for carbonate, formate and oxalate in the spectrum of HEDTA-B were very intense compared to those for HEDTA-A. This indicates that the carboxymethyl groups are also major sources of these products. Ion chromatographic analyses of the HEDTA-B product mixture also indicated the presence of formate (111%), oxalate (22%) and glycolate (27%). Workers at PNL earlier reported (25) the formation of formate and glycolate in an irradiated sample containing sodium citrate, sodium EDTA, sodium HEDTA and inorganics.

The Mechanism of HEDTA Degradation in Simulated Waste (29)

Kinetic data, evidence for formaldehyde as an intermediate, demonstration of the decomposition of the nitrite ester of glycolate to produce the same products as when glycolate decomposes in simulated waste solution, and the ability to identify, by ^{13}C-NMR, at least ten by-products from HEDTA decomposition, provides many clues as to the mechanism of decomposition of HEDTA in simulated waste. The mechanism proposed in Scheme 4 is consistent with the data currently available.

The first two reactions of Scheme 4 are key to the understanding of how HEDTA decomposition proceeds. In the first reaction nitrite ion (NO$_2^-$) reacts with aluminate (Al(OH)$_4^-$) to substitute one of the hydroxide ligands; the coordinated nitrito ligand is expected to be much more electrophilic and thus more reactive toward HEDTA, which

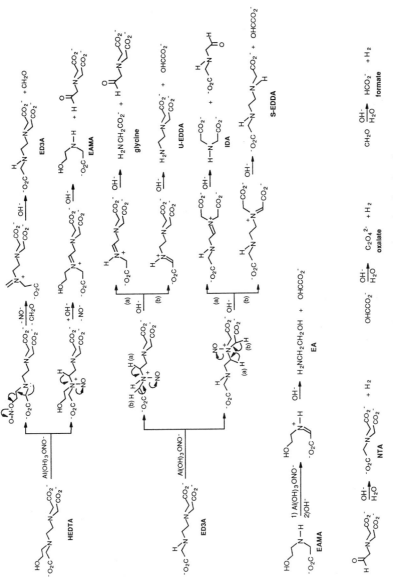

Scheme 4. Mechanism of HEDTA reactions in simulated waste.

has several nucleophilic sites. In the second reaction the activated nitrito species acts as an NO^+ source to react with HEDTA. Two intermediates, one produced by reaction of the HEDTA hydroxyl group and the other by reaction of the tertiary nitrogen to which the hydroxyethyl group is bonded, and their subsequent reactions are shown. In principle the other tertiary nitrogen center could also react with the activated nitrite. However, the failure of EDTA to react with nitrite under identical conditions suggests that for some reason this may not be a viable pathway.

Using model compounds we have shown that a $-CH_2OH$ group can be converted to the $-CH_2ONO$ group and that this group collapses to formaldehyde and NO^- with formaldehyde reacting further to generate formate and H_2. In the present case, the imine that results from fragmentation of the nitrite ester undergoes base hydrolysis to form more formaldehyde and ED3A. NMR studies have shown that ED3A is a major intermediate in the HEDTA decomposition.

That there is some reaction at the tertiary nitrogen is strongly suggested by the identification of EAMA. Formation of this product requires cleavage of the ethylene diamine backbone of the HEDTA molecule while the hydroxyethyl group remains intact. Production of ethanol amine is shown as arising from further reaction of EAMA with the nitrito aluminate species latter in the Scheme.

Further reaction of ED3A, by a similar series of steps to that of HEDTA, can result in the formation of IDA, S-EDDA, U-EDDA and glycine, all of which were identified by ^{13}C NMR as products of HEDTA decomposition. We have shown attack of ED3A by NO^+ at the tertiary center, whereas as mentioned above this may not be a viable pathway for HEDTA (or EDTA). Such a process is necessary to account for the formation of S-EDDA, which is clearly present in the product mixture from HEDTA and which we know from independent experiments is produced from ED3A. We believe that the answer to this apparent inconsistency lies in the fact that ED3A can form a lactam which results in a substantial increase in the acidity of the protons of one of the cyclized carboxymethyl groups. This would make the following process feasible.

Alternatively, attack of $[NO]^+$ on the secondary amine followed by *intramolecular* transfer to the tertiary center to give the intermediate shown might also be possible. For the sake of simplicity neither of these modifications have been incorporated into Scheme 4. Further degradation of these products generates formate, oxalate, CO_2 and H_2.

Scheme 4 does not incorporate the formation of N_2O, N_2 and NH_3. As described earlier in this paper, these species arise predominantly from the inorganic constituents of the waste mixture. We are continuing to pursue the details of their formation and expect to present a detailed reaction sequence for their formation in a subsequent report.

Effect of Oxygen on the Reactions of HEDTA and Related Compounds in Simulated Waste

Table XII shows the reactivities for EDTA, Glycolate, HEDTA and its fragments, under argon and oxygen atmospheres. All these reactions were carried out in the apparatus shown in Figure 4 provided with the Teflon cap shown in Figure 12. In all cases the reactivities are enhanced under oxygen atmosphere. Very little or no reaction is observed under argon atmosphere for all compounds that do not contain a hydroxyethyl function, an exception being ED3A-lactam. Although ED3A-lactam undergoes rapid ring opening in simulated waste solution at 120° C, it is possible that some kind of chain reaction is initiated prior to the ring opening. Preliminary results indicate that S-EDDA lactam (which also undergoes rapid ring opening under the conditions mentioned above) reacts much faster than S-EDDA. These results could explain why ED3A is observed as a major product in the reaction of HEDTA in simulated waste solution although its lactam reacts rapidly; the open form reacts much slower than the lactam, and therefore its concentration builds up. Based on the amounts of starting material recovered, an order of reactivities could be established:

Glycolate > HEDTA, ED3A(lactam) > > U-EDDA, S-EDDA > EDTA, glycine, IDA, NTA

Sodium glycolate is by far the best hydrogen producer, and is also the best $N_2 + N_2O$ producer, sharing this property with HEDTA. It is clear from the observed reactivities that those compounds with hydroxyethyl groups are much more reactive than those without hydroxyethyl units (except the lactams). With respect to gas generation, in practically all cases, the amount of hydrogen produced is substantially larger under oxygen than under argon atmosphere. This result should be considered in evaluating possible mitigation procedures. Procedures which introduce O_2 into the waste mixture should increase the rate of H_2 formation.

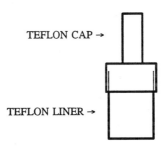

TEFLON CAP →

TEFLON LINER →

Figure 12. Teflon Liner with Cap.

Table XII. Reaction of EDTA, Glycolate, HEDTA, and HEDTA Fragments in Simulated Waste Solution (120° C, 1000 h)[a]

Trial Name	Atm.[d]	%Starting Material	%HCOO⁻	%C₂O₄⁼	%H₂	%O₂[b]	%N₂[c]	%CH₄	%N₂O
EDTA-5A	Ar	102	<1	<2	1.3	-	~4	<0.1	<0.1
Glycolate-14A	Ar	27.4	29.8	43.5	21.2	-	~7	0.3	14.6
HEDTA-22A	Ar	34.8	73.4	10.0	2.7	-	20	0.1	5.9
ED3A-1A[4]	Ar	43	6.4	5.1	1.1	-	~5	0.1	1.1
Glycine-1A	Ar	97	1.4	6.2	1.5	-	~4	<0.1	0.4
u-EDDA-1A	Ar	90[f]	7.0	<2	0.5	-	~7	0.1	0.5
s-EDDA-2A	Ar	86	1.9	<2	2.0	-	~10	0.8	0.5
IDA-1A	Ar	100	<1	<2	0.9	-	~3	<0.1	<0.1
NTA-2A	Ar	95	1.9	5.0	2.5	-	~11	0.2	<0.1
EDTA-6A	O₂	93	3.0	9.7	6.7	17	~3	0.1	<0.1
Glycolate-12A	O₂	7.8	34.4	48.3	26.8	25	~5	0.3	9.5
HEDTA-21A	O₂	30.4	79.9	10.5	2.6	24	~11	0.3	8.6
ED3A-2A[e]	O₂	24	13.3	10	8.1	31	~5	0.3	1.3
Glycine-1A	O₂	81	2.0	16.5	8.5	15	~3	<0.1	0.1
u-EDDA-2A	O₂	80[f]	17.8	4.8	2.2	23	13	<0.1	0.1
s-EDDA-1A	O₂	79	7	15.3	9.3	25	~6	0.1	0.3
IDA-2A	O₂	93	2.3	4.9	4.1	15	~8	<0.1	0.1
NTA-1A	O₂	92	3.3	11.6	6.7	22	~2	0.1	<0.1

[a]Unless otherwise stated, all percentages represent moles of product per 100 moles of organic starting material. [b]Amount of oxygen *consumed* per 100 moles of starting organic material. [c]A considerable deviation (~ 40% when %N₂ is < 10) can be expected for these values, due to air contamination. [d]The gas pressure above the solution was ~350mm. [e]The compound was added in the forms of its lactam, which under those conditions undergoes rapid ring opening. [f]Estimated values.

Acknowledgment: We are indebted to the Westinghouse Hanford Company for support of this work.

Literature Cited

(1) Delegard, C. *RHO-LD-125*, Dec. **1980**; Avail. INIS, NTIS.

(2) Lokken, R. O; Scheele, R. D; Strachan, D. M; Toste, A. P. *PNL-7687*, Sept. **1986**; Avail. INIS, NTIS.

(3) Bhattacharyya, S. N; Kundu, K. P. *Int. J. Radiat. Phys. Chem.*, **1972**, *4*, 31.

(4) Bhattacharyya, S. N; Srisankar, E. V. *Int. J. Radiat. Phys. Chem.*, **1976**, *8*, 667.

(5) Bhattacharyya, S. N; Saha, N. C. *Radiation Res.*, **1976**, *68*, 234.

(6) Hammett, L. P. *Physical Organic Chemistry*, McGraw-Hill, New York, **1940**, p. 350.

(7) Loew, O. *Ber.*, **1887**, *20*, 144.

(8) Siemer, D. D., reported by Strachan, D. M., in *Minutes of the Tank Waste Science Panel Meeting*, Feb. 7-8, **1991**, *PNL-7709.*

(9) Activities were calculated from data in Akerlof, G; Kegeles, G. *J. Am. Chem. Soc.*, **1940**, *62*, 620.

(10) Gruen, L. C; Mc Tigue, P. T. *J. Chem. Soc.*, **1963**, 5217.

(11) Von Euler, H; Lovgren, T. *Z. Anorg. Allgem. Chem.*, **1925**, *147*, 123.

(12) Bell, R. P; Onwood, D. P. *Trans. Faraday Soc.*, **1962**, *58*, 1557.

(13) Wilmarth, W. K; Dayton, J. C; Flournoy, J. M. *J. Am. Chem. Soc.*, **1953**, *75*, 4549; Flournoy, J. M; Wilmarth, W. K. *J. Am. Chem. Soc.*, **1969**, *83*, 227.

(14) Walker, J. F. *Formaldehyde*. Reinhold Publishing Corp., 2nd ed., **1953**, 187.

(15) Wieland, H.; Wingler, A. *Ann.* **1923**, *431*, 301.

(16) Wirtz, K; Bonhoeffer, K. F. *Z. Physik. Chem.*, **1936**, *B32*, 108.

(17) Jaillet, J. B.; Quellet, C. *Can. J. Chem.*, **1951**, *29*, 1046.

(18) Abel, E. *Z. Physik. Chem. N. F.*, **1956**, *7*, 101.

(19) Hartwagner, F. *Z. Anal. Chem.*, **1913**, *52*, 17.

(20) Bonner, F. T.; Hughes, M. N. *Comments Inorg. Chem.*, **1988**, *7*, 215, and references therein.

(21) Doyle, M. P; Terpstra, J. W; Pickering, R. A; Le Poire, D. M. *J. Org. Chem.*, **1983**, *48*, 3379.

(22) McLendon, G; Motekaitis, R. J; Martell, A. E. *Inorg, Chem.*, **1975**, *14*, 1993.

(23) Rowley, G. L.; Greenleaf, A. L.; Kenyon, G. L. *J. Am. Chem. Soc.*, **1971**, *93*, 5542.

(24) Genik-Sas Berezowsky, R. M; Spinner, I. H. *Can. J. Chem.*, **1970**, *48*, 163.

(25) Campbell, J. A; Pool, K. H; Melethil, P. K. *PNL-8041*, March **1992**.

(26) Initial disclosure of this work can be found in, Strachan, D. M., *Hanford Tank Safety Project, Minutes of the Tank Waste Science Panel Meeting*, February 7-8, **1991** (*PNL-7709, AD-940*, June 1991).

(27) Initial disclosure of this work can be found in, Strachan, D. M., *Hanford Tank Safety Project, Minutes of the Tank Waste Science Panel Meeting*, July 9-11, **1991** (*PNL-8048*, April 1992).

(28) Initial disclosure of this work can be found in, Strachan, D. M., *Hanford Tank Safety Project, Minutes of the Tank Waste Science Panel Meeting*, November 11-13, **1991** (*PNL-8047*, April 1992).

(29) Initial disclosure of this work can be found in, Strachan, D. M., *Hanford Tank Safety Project, Minutes of the Tank Waste Science Panel Meeting*, March 25-27, **1992** (*PNL-8278, AD-940*, August 1992).

(30) Since this symposium presentation, Dr. Dan Meisel of Argonne National Laboratory has indicated (personal communication) that deuterium labeling experiments of formaldehyde and water show that, under their experimental conditions (0.13 M CH_2O, 2.3 M NaOH, 23˚C), a large fraction of the H_2 produced comes from the hydrogen atoms on formaldehyde. Our results indicate that at these conditions, the maximum H_2 yield, formed in competition with the Cannizzaro reaction, would be about 0.1 %.

RECEIVED December 1, 1993

Chapter 17

Factors Affecting the Rate of Hydrolysis of Phenylboronic Acid in Laboratory-Scale Precipitate Reactor Studies

Christopher J. Bannochie, James C. Marek,
Russell E. Eibling, and Mark A. Baich

Westinghouse Savannah River Company, Savannah River Technology
Center, Aiken, SC 29802

Removing aromatic carbon from an aqueous slurry of cesium-137 and other alkali tetraphenylborates by acid hydrolysis will be an important step in preparing high-level radioactive waste for vitrification at the Savannah River Site's Defense Waste Processing Facility (DWPF). Kinetic data obtained in bench-scale precipitate hydrolysis reactors suggest changes in operating parameters to improve product quality in the future plant-scale radioactive operation. The rate-determining step is the removal of the fourth phenyl group, i.e. hydrolysis of phenylboronic acid. Efforts to maximize this rate have established the importance of several factors in the system, including the ratio of copper(II) catalyst to formic acid, the presence of nitrite ion, reactions of diphenylmercury, and the purge gas employed in the system.

The Defense Waste Processing Facility (DWPF) at the Savannah River Site is a waste vitrification facility designed to immobilize high-level radioactive isotopes in borosilicate glass. The radionuclides are composed primarily of cesium-137 but include strontium-90, uranium and plutonium, as well as other trace fission products, currently stored as salt-cake and sludge in underground tanks. The facility is composed of three main processing elements: a Salt Cell, a Chemical Cell, and a Glass Melter Cell. This paper will deal with laboratory-scale work performed in support of operation of the Salt Processing Cell.

Sodium tetraphenylborate is used to precipitate cesium-137 from the tank farm salt waste while monosodium titanate is added to adsorb the strontium-90 and isotopes of uranium and plutonium. An insoluble tetraphenylborate salt is formed with cesium-137, as well as with potassium and ammonium ions also present in the waste. The insoluble tetraphenylborate salts will be fed to the DWPF Salt Processing Cell for hydrolysis and removal of aromatic carbon. Removal of aromatic carbon is essential to facilitate the melter operation and assure glass quality. The resulting aqueous product (PHA) is then transferred to the Chemical Processing Cell for concentration and combination with sludge and frit prior to transfer to the glass melter for vitrification.

Bench-scale precipitate hydrolysis reactors in conjunction with both unirradiated and irradiated, simulated waste, have been utilized to study the processing of this material for future plant-scale radioactive operation. The hydrolysis is carried

0097–6156/94/0554–0285$08.00/0
© 1994 American Chemical Society

out with formic acid and copper(II) as catalyst to yield primarily benzene and boric acid.

$$(C_6H_5)_4BX + H^+ \longrightarrow C_6H_6 + (C_6H_5)_3B + X^+ \tag{1}$$

$$(C_6H_5)_3B \xrightarrow[\text{H}_2\text{O}]{\text{H}^+ , \text{Cu(II)}} C_6H_6 + (C_6H_5)_2B(OH) \tag{2}$$

$$(C_6H_5)_2B(OH) \xrightarrow[\text{H}_2\text{O}]{\text{H}^+ , \text{Cu(II)}} C_6H_6 + (C_6H_5)B(OH)_2 \tag{3}$$

$$(C_6H_5)B(OH)_2 \xrightarrow[\text{H}_2\text{O}]{\text{H}^+ , \text{Cu(II)}} C_6H_6 + B(OH)_3 \tag{4}$$

The reactions shown in equations 1-4 are believed to proceed by way of a free-radical mechanism by analogy with aromatic diazonium chemistry. The Sandmeyer (Cu(I) catalyzed) and Gattermann (Cu(0) catalyzed) reactions which yield arylhalides from aryldiazo compounds as well as the Gomberg reaction which gives substituted biaryls from diazohydroxides and neutral aromatic compounds in aqueous systems (equation 5) have been shown to proceed by way of free-radical mechanism (1,2).

$$ArN_2OH + C_6H_5R \longrightarrow ArC_6H_4R + N_2 + H_2O \tag{5}$$

The by-products of these reactions include biphenyl and phenol which are also seen in our system. Kuivila and co-workers (3-5) did not attribute a free-radical nature of this reaction in their kinetic studies on areneboronic acids. However, they did recognize the unique role of copper(II) in the catalysis of these reactions (5). Kinetic data obtained here suggested changes in operating parameters to improve the quality of the final products. Several factors have been identified which affect the rate of hydrolysis of phenylboronic acid (PBA). These factors include the ratio of copper(II) to formic acid, the presence of nitrite ion, diphenylmercury, and complexing anions, as well as the nature of the purge gas.

Experimental

Materials. All Metal salts used in the preparation of precipitate feeds were reagent grade. The sodium tetraphenylborate solution was obtained from Optima Chemicals (Tank #730-A). Analysis of this material is shown in Table I. Copper(II) formate was purchased from Sheppard Chemical (Lot B49860) and was found to have 33.1 ± 0.2 wt% Cu (essentially $Cu(HCO_2)_2 \cdot 2H_2O$, which is 33.5 wt% Cu). Formaldehyde solutions were 90 wt% reagent grade obtained from Fisher Chemical. Deionized water was used to prepare all solutions.

Two types of non-radioactive precipitate slurries were studied: 1) an unirradiated precipitate slurry that simulates the projected composition of the plant's radioactive precipitate after two years of storage and 2) an irradiated precipitate slurry that was prepared to simulate the Savannah River Site's (SRS) freshly prepared radioactive slurry and was subsequently exposed to gamma radiation from a cobalt-60 source. The irradiated slurries were given an absorbed dose of 1.9E+08 rads by exposing them to a gamma dose rate of 3.5E+06 rads/hr for 55 hours.

TABLE I

Analysis of Sodium Tetraphenylborate Solution

Component	Amount Detected
TPB⁻	16.76 wt%
OH⁻	0.18M
phenylboronic acid	552 ppm
N-phenylformamide	——
aniline	19 ppm
phenol	173 ppm
nitrobenzene	——
nitrosobenzene	——
4-phenylphenol	——
2-phenylphenol	7 ppm
diphenylamine	22 ppm
biphenyl	126 ppm
o-terphenyl	——
m-terphenyl	——
p-terphenyl	——

Both types of precipitate slurries were prepared by adding the vendor-supplied solution of sodium tetraphenylborate to a solution of potassium salts containing the monosodium titanate and sodium nitrite (and diphenylmercury, when appropriate). The unirradiated precipitate simulant was prepared and sampled for analysis of composition prior to each experiment. The salts were added based on the average blend of salts from the material balance for SRS salt waste. Table II shows the composition of the unirradiated precipitate simulant and the composition of the second precipitate simulant before irradiation. The slurry for irradiation contained diphenylmercury at three times the nominal flowsheet level and initially, about 0.25M nitrite. The irradiated slurry, after removal from the cobalt well, was analyzed for organics and inorganics. It was then adjusted to 0.08M nitrite and washed in a continuous washing operation to reduce the nitrite concentration to 0.01M. The washing was performed by pumping the irradiated slurry through a sintered-metal crossflow filter and adding wash water to the agitated slurry at the same rate at which filter permeate was withdrawn.

Equipment and Hydrolysis Conditions. Hydrolysis experiments were performed in a 2 L cylindrical glass vessel with a tapered flat bottom fitted with a temperature probe, stainless steel dual 4-blade mechanical stirrer, stainless steel feed tube, glass sampling tube, and insulated glass vapor line connected to a condenser followed by a glass decanter. Aqueous condensate returned from the decanter to the reactor vessel. Experiments were also conducted in a 1 L Hastelloy C-276 vessel to simulate the plant reactor vessel (no difference was noted in the PBA kinetics from that observed with the glass reactor vessel). A Cole-Palmer Instruments Masterflex pump (Model 7550-90) was employed to feed the precipitate slurry into the reactor vessel below the heel level over a two-hour period. The 100 grams of deionized water used to rinse the feed lines was approximately equivalent to the decanter aqueous holdup. Temperature was maintained at 90°C with a Cole-Palmer Dyna-Sense

Electronic Temperature Controller (model 2155) during the feed period and for an additional five hours post-feed. During this period, 20 g samples were withdrawn every hour, but occasionally more frequently as required to monitor the PBA, copper,

TABLE II

Composition of Unirradiated Precipitate Simulant and Irradiated Precipitate Simulant Prior to Irradiation

Component	Unirradiated (wt. %)	Irradiated (wt. %)
H_2O	89.2	87.8
$KBPh_4$	8.10	8.86
$CsBPh_4$	0.113	0.111
$NaBPh_4$	0	0.537
$NaNO_3$	0.0212	0.171
$NaNO_2$	*	1.56
Na_2SO_4	0.0203	0.0205
Na_2CO_3	1.87	0.0189
$NaOH$	0.15	0.363
Na_2SiO_3	0.000474	0.000481
$NaCl$	0.00270	0.00135
NaF	0.000756	0.000646
Na_2CrO_4	0.000543	0.000552
K_2CO_3	0.164	0
$CsOH$	0	1.55×10^{-10}
$Na_2B_4O_7$	0.0402	0.0224
C_6H_5ONa	0.151	
	0.000459	
$Hg(C_6H_5)_2$	0	0.315
$NaTi_2O_5H$	0.200	0.200

* Added into simulant at levels anticipated after processing, see text.

and diphenylmercury (when present) levels. Following the reaction period at 90°C, the temperature was increased to 102°C for an additional five hours for aqueous stripping of the high-boiling organic by-products. In the decanter, the condensed aqueous phase drops through the benzene organic phase, which collected during the 90°C reaction cycle, transferring the dissolved organic by-products such as biphenyl and diphenylamine. During the entire reaction period the system is kept under a carbon dioxide purge at a flow rate of 2.8 std. cm^3/min.

The reaction heel is 412.5 g and is scaled to duplicate the one necessary to cover the steam and cooling coils in the pilot-scale facility. The heel contains formic acid, copper formate, and deionized water. It is charged to the reactor vessel and brought to 90°C prior to the addition of the precipitate slurry.

Analytical Measurements. All samples were analyzed by reverse phase, gradient liquid chromatography using a water/acetonitrile solvent system. The analytical column was a 25 cm, 4.6 mm ID Chemcosorb 5-ODS-UH (30% carbon load, bonded phase C-18). The samples were run in a Hewlett Packard 1090 with a diode array detector. Phenylboronic acid was analyzed at 217 nm while diphenylmercury was analyzed using 225 nm.

Standards were prepared from reagent grade chemicals without further purification. Quantitation was based upon external standards. Sample preparation consisted of dilution with "HPLC Grade" acetonitrile and filtration through a 0.2 micron nylon filter. Samples with pH greater than 8 were adjusted with reagent grade potassium dihydrogen phosphate prior to dilution to prevent damage to the silica based C-18 column.

A Varian SpectrAA 3400 Atomic Absorption Spectrophotometer was employed in the determination of Cu(II). Standardization was by dilute (2, 4, and 6 ppm) NBS standards. Aqueous samples were prepared with 0.3M HNO_3 to insure sufficient acidity upon dilution.

Results and Discussion

In order to examine the effects of the copper:acid ratio, nitrite, purge gas, and complexones, the unirradiated, precipitate simulant was employed. The desired level of nitrite was added and a corresponding change in the amount of water was made. This precipitate slurry simulated certain features of the radiolyzed slurry (sodium, carbonate, phenol) but many of the radiolysis products (benzene, biphenyl, phenylboronic acid, phenylphenols, terphenyls, formate, oxalate) were left out of the basic recipe to determine whether these materials are produced by hydrolysis reactions. Radiolysis will decompose some of the potassium and cesium tetraphenylborate during storage. The precipitate produced during the first 5 years of waste processing is projected to have an average activity level of 24 Ci/gal cesium-137. Over the life of the facility, the activity level is projected to average 36 Ci/gal cesium-137. During two years of storage, the precipitate will absorb doses of 1.9E+08 and 2.9E+08 rads, respectively, at 24 Ci/gal and 36 Ci/gal cesium-137. The effect of diphenylmercury was examined by utilizing the washed, irradiated precipitate slurry described above.

Ratio of Copper(II) to Total Acid. In a series of experiments, unirradiated simulant was hydrolyzed under a series of total acid concentrations at each of three copper(II) formate catalyst levels: 475, 950, and 1425 ppm. In each of these experiments only the amount of initial formic acid in the pre-reaction heel was varied. The level of phenylboronic acid (PBA) was followed by HPLC in samples taken at one hour intervals throughout the five-hour reaction period at 90°C and aqueous boil period at 102°C. Additional samples were drawn during the first hour of aqueous boil when necessitated by the rate of loss of PBA. Since loss of the final phenyl ring is rate-determining in the hydrolysis of tetraphenylborate, the PBA levels allowed us to judge the degree of completion of the reaction. A plot of the natural logarithm of [PBA] versus time gave a straight line consistent with a first order loss of PBA. The slope of this line is reported here as an observed first order rate constant. Figure 1 provides an example of the type of correlations observed in the data. At 475 ppm Cu(II), it was not possible to eliminate the PBA during the reaction period at any of the acid levels examined. Figure 2 indicates the levels of PBA remaining in the reactor solution at the end of the five-hour reaction period at 90°C. The reaction temperature of 90°C was chosen after consideration of the effect of diphenylmercury on the reaction system (to be discussed below). In fact, at 475 ppm Cu(II) it was not possible to eliminate the PBA, even after five additional hours at 102°C. Currently, the process limit for PBA in the final aqueous product is 53 ppm. When the level of Cu(II) was increased to 950 ppm there was a significant increase in the rate of PBA hydrolysis, and the levels after five hours of reaction at 90°C ranged from 300 - 500

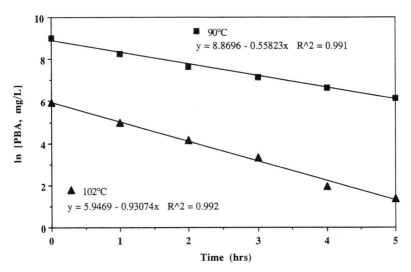

Figure 1.   Plot of the ln[PBA] versus time (hrs) for an unirradiated, tetraphenylborate precipitate slurry without nitrite ion with 950 ppm Cu(II) and 0.3M total acid at 90°C followed by five additional hours at 102°C.

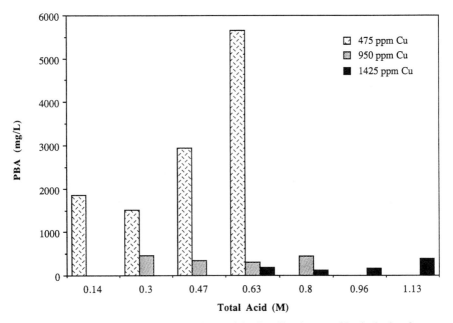

Figure 2. Phenylboronic acid levels (mg/L) after five hours of hydrolysis of an unirradiated, tetraphenylborate precipitate without nitrite ion at various level of Cu(II) as a function of total acid (M) at 90°C.

ppm with complete hydrolysis achieved by the completion of the aqueous stripping period. At 1425 ppm Cu(II) the levels were reduced even further, to between 125 - 400 ppm, during the initial five hours at 90°C.

The most striking observation concerned the relationship between the copper(II) formate catalyst and the amount of total acid necessary to maximize the rate of hydrolysis of PBA. At each successive catalyst level, additional formic acid was necessary to maximize the rate of hydrolysis. Even more striking was the parabolic relationship between the Cu(II) and the total titratable acid in the system. Figure 3 shows this relationship for systems both with and without nitrite ion (to be discussed below). With either too little or too much acid, there was a decrease in the observed rate constant for hydrolysis of PBA. It was also noted that there was a correlation between the maximum observed rate constants for each Cu(II) level.

Effect of Nitrite Ion. Sodium nitrite is added to the waste tanks to inhibit pitting of the carbon steel. The level of nitrite added depends on the amount of nitrate and hydroxide present, as well as on the temperature of the tank waste. The irradiation process reacted over 90% of the nitrate and nitrite ions in the simulant, converting the nitrite ions to primarily nitrous oxide (some of the nitrite ions also form ammonia and/or ammonium ions, as well as nitroso and nitro aromatics). In SRS waste processing, periodic nitrite ion additions will be required to replace nitrite ions radiolytically destroyed. Our simulant contained sufficient nitrite ion that 300-400 ppm remained after the irradiation process. The amount of nitrite ion expected in the DWPF Salt Processing Cell will depend on the final upstream processing of the precipitate slurry. The anticipated amount was simulated with nitrite ion adjustment and slurry washing.

In this work, the effect of 0.01M nitrite ion under the conditions identified to give the maximum rate of PBA hydrolysis for each Cu(II) level were examined. This is the level of nitrite ion expected if the precipitate slurry is washed. Again there was a correlation between the observed rate constants for the three levels of catalyst examined. When various levels of formic acid were utililized to affect a change in the total acid level of the 950 ppm Cu(II) system, there was more flexibility in the ratio of acid in the presence of nitrite ions. The acid level could be reduced from 0.6 M to 0.3 M and maintain essentially the same observed rate of hydrolysis of PBA. However, below 0.3 M total acid there was a significant decrease in the observed rate constant (Figure 3).

To further examine the effect of nitrite ion on the PBA kinetics, the starting feed level in one experiment was increased to 0.02M and decreased to 0.005M in another. The results of these changes on the rate constant for PBA hydrolysis are shown in Figure 4. As anticipated, the changes in the rate constant were directly proportional to the nitrite ion level in the feed. Nitrite ion may be involved in the catalytic cycle of the copper catalyst, possibly serving to oxidize Cu(I) back to Cu(II) in much the same manner that occurs in the conversion of nitrous acid to nitric oxide by Fe(II) (*6*). Nitrous acid with a pK_a of 3.0 (25°C, $I = 1.0$) (*7*) is readily formed in this buffered system, pH 3.5 - 4.0.

$$Fe(II) + HNO_2 + H^+ \longrightarrow Fe(III) + NO + H_2O \qquad (6)$$

$$Cu(I) + HNO_2 + H^+ \longrightarrow Cu(II) + NO + H_2O \qquad (7)$$

Equation 7 has an even more positive standard cell potential ($E^0 = 0.83$ V) than equation 6, $E^0 = 0.23$V, and thus would also be expected to occur spontaneously. It has been shown that PBA when refluxed with cupric chloride produces cuprous choride (*8*).

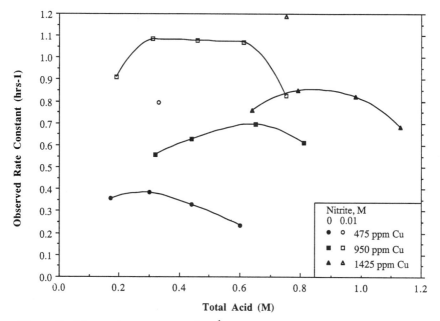

Figure 3. Observed rate constant (hrs^{-1}) for hydrolysis of phenylboronic acid versus total acid (M) for varying levels of Cu(II) at 90°C both with (0.01 M) and without nitrite ion in an unirradiated, tetraphenylborate precipitate.

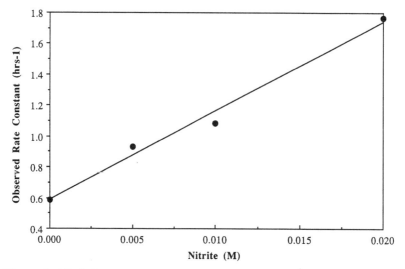

Figure 4. Variation in the observed rate constant (hrs^{-1}) for hydrolysis of phenylboronic acid as a function of the initial sodium nitrite level (M) in an unirradiated, tetraphenylborate precipitate.

    While the nitrite ion has a positive effect on the hydrolysis of PBA, it has a detrimental effect on the levels of by-products produced through reactions of phenyl radicals with nitrous acid. These by-products include diphenylamine, aniline, N-phenylformamide, nitrobenzene, nitrosobenzene, and possibly other nitrogen containing byproducts not yet identified. Amine containing compounds such as aniline result from reduction of nitrosobenzene. As a free-radical trap, nitrosobenzene can react with phenyl radicals to produce diphenylnitroxide which can undergo reduction to diphenylamine. The aqueous boiling period is designed to steam strip the high-boiling organic compounds which result from both radiolysis of the stored precipitate slurry as well as from by-products of precipitate hydrolysis. Since the level of diphenylamine is directly related to the starting nitrite ion level of the feed (Figure 5) and the difficulty of steam stripping this material from the aqueous product due to its low vapor pressure (b.p. 302°C), it is preferable to minimize the level of nitrite ion in the tetraphenylborate slurry prior to precipitate hydrolysis.

Effect of Purge Gas. The DWPF Salt Processing Cell employs CO_2 as the inertant gas. The fact that nitrite ion contributes to an increase in the rate of PBA hydrolysis lead us to look at the effect of substituting air, and hence oxygen, into the system. In an experiment where nitrite ion was left out of the simulant feed and air was used as the purge gas over the system, though it was not sparged through the reaction solution, an increase in the PBA hydrolysis rate constant was obtained which matched that observed in the presence of nitrite ion (Table III). These observations support the belief that factors which can influence the catalytic cycle of the copper catalyst affect the rate of hydrolysis of PBA.

$$4Cu(I) + O_2 + 4H^+ \longrightarrow 4Cu(II) + 2H_2O \qquad (8)$$

At pH 3.5 and 90°C, assuming unit activity for oxygen and equal amounts of Cu(I) and Cu(II) in the system, the Nernst Eqn. gives an $E_{cell} = 0.30$ V for equation 8, indicating a spontaneous reaction is possible.

TABLE III

Influence of Nitrite Ion and Purge Gas on the Observed Rate Constant for Hydrolysis of Phenylboronic Acid

Purge Gas	$[NO_2^-]$ (M)	$[Cu(II)]_{initial}$ (ppm)	Total Acid (M)	Obs. Rate Const. (hrs^{-1})
CO_2	0	475	0.30	0.38
CO_2	0.01	475	0.32	0.67
air	0	475	0.32	0.68

Complexation of Catalyst. Several anionic species are anticipated in the DWPF feed, either from the initial waste stream or as a result of radiolysis, including phosphate and oxalate. Radiolysis produces currently unidentified ionic species which, as previously noted, are probably one or more organic acids. These species have been shown to inhibit the catalyst when not removed by washing from the irradiated simulant prior to hydrolysis. To further examine the effect of complexones an experiment was run wherein the Cu(II) was complexed in the pre-reaction heel with N,N'-ethylenediaminetetraacetic acid (EDTA). All other variables were constant, i.e.

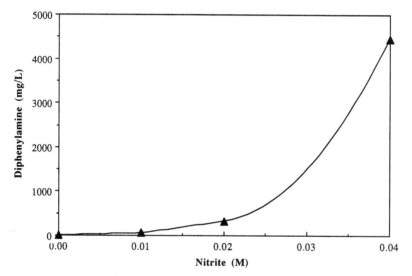

Figure 5. Variation in the level of diphenylamine remaining in the aqueous product from an unirradiated, tetraphenylborate precipitate after five hours of aqueous stripping at 102°C as a function of initial sodium nitrite level (M).

total acid 0.3M, 950 ppm Cu(II), 0.01M NO_2^-. The PBA levels were monitored as in previous runs and varied from 15,500 mg/L, at the start of the 90°C reaction period, to 14,900 mg/L after five hours. At the end of the five hours of aqueous stripping the PBA levels had only fallen to 13,200 mg/L. It is apparent that EDTA is able to prevent the Cu(II) catalyst from successfully initiating the free-radical mechanism.

When phosphate was added into an unirradiated precipitate slurry at a level corresponding to 1.1 wt%, there was an approximately 20% decrease in the PBA hydrolysis rate constant from 1.1 hrs^{-1} to 0.91 hrs^{-1}. A layer of blue solids was also formed in the reactor vessel. These solids were identified by Scanning Electron Microscopy (SEM) to contain primarily Cu_3PO_4, a highly insoluble copper(II) salt. By precipitating copper from the system, phosphate had a detrimental impact on the hydrolysis of PBA. Hence it appears that strong complexing agents with an affinity for Cu(II) can interfere in the mechanism of hydrolysis of PBA.

Effect of Mercury. Diphenylmercury is formed during the precipitation of cesium-137 through reaction of mercury present in the waste tanks with tetraphenylborate. Hu and Chang (9) proposed the following reaction between mercury and $NaBPh_4$ in base:

$$Hg + [B(C_6H_5)_4]^- + OH^- \longrightarrow Hg(C_6H_5)_2 + (C_6H_5)_2BOH + 2e^- \quad (9)$$

While Heyrovsky (10) showed that $PhHg^+$ forms completely prior to the separation of any $KBPh_4$.

The five hour reaction time at 90°C was brought about initially by an examination of the effect of Ph_2Hg on precipitate hydrolysis. When precipitate slurry was fed to the reactor vessel over the standard two hour period at 90°C and followed immediately by aqueous stripping at 102°C, Ph_2Hg levels in the decanter organic product were in the range 1000-1500 mg/L. The current disposal limit on mercury in the organic product is 260 ppm. In an attempt to hydrolyze the Ph_2Hg more rapidly to Hg(II), which is readily reduced to Hg(0) by formic acid and does not collect in the organic phase, 98°C feeding was initiated. Unfortunately, the Ph_2Hg levels in the organic product increased to 2000-3000 mg/L. Figure 6 presents a semilogarithmic plot of vapor pressure (P) versus 1/T for various mercuric halides. Assuming that the slope for the ln(P) versus 1/T for diphenylmercury is similar to these compounds, an estimate can be made for the vapor pressure of Ph_2Hg using the data (11) for HgI_2.

$$\ln (P)_{Ph_2Hg} \approx \ln (P)_{HgI_2} = 23.38 - 1.007E4 / T \quad (10)$$

The estimate indicates an increase by a factor of two on going from 90°C to 98°C. So in order to hydrolyze the Ph_2Hg and prevent it from steam stripping into the organic product, the Ph_2Hg levels were examined in the organic product after 90°C reaction periods of two, four and five hours. At five hours, three times the nominal anticipated feed level of Ph_2Hg was hydrolyzed successfully resulting in no Ph_2Hg in the organic product.

When the washed, irradiated, precipitate simulant was sampled and analyzed for Ph_2Hg during measurement of PBA levels some interesting observations were made (Figure 7). When Ph_2Hg was present in the slurry there was no longer a first-order loss of PBA. Additionally the level of Ph_2Hg remained essentially constant until the level of PBA dropped to zero. Only upon complete hydrolysis of PBA was it possible to begin to hydrolyze the Ph_2Hg. These observations lead us to speculate that any phenylmercurate ion produced, reacted with PBA to regenerate Ph_2Hg, thus

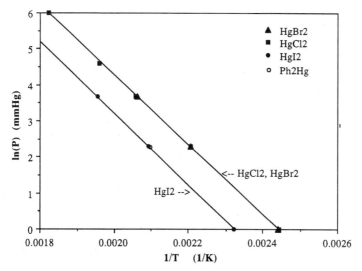

Figure 6. Variation in vapor pressure, ln(P) (mmHg), of various mercuric halides as a function of temperature, 1/T (1/K), Also shown is the single datum point for diphenylmercury [10.5 mmHg at 204°C].

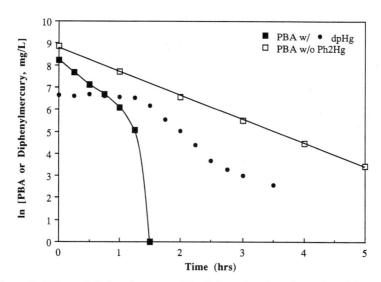

Figure 7. Effect of diphenylmercury (Ph₂Hg) on the phenylboronic acid (PBA) hydrolysis kinetics at 90°C with 950 ppm Cu(II), 0.3M total acid, and 0.01M sodium nitrite. The change in Ph₂Hg concentration in shown by the unconnected solid circles.

maintaining a constant level of Ph_2Hg while providing a second catalytic mechanism for hydrolysis of PBA.

Conclusions. The rate of hydrolysis of phenylboronic acid (PBA) is influenced by several factors. Maximization of the rate is achieved by balancing the ratio of Cu(II) and total acid (formic acid) in the system. This ratio is significantly affected by the presence or absence of nitrite ion in the initial tetraphenylborate precipitate, with nitrite ion serving to increase the observed rate and make it less dependent on the level of acid in the system. The presence of oxygen in the system also increases the observed rate of hydrolysis. Complexing agents which interfere with the catalytic cycle of the Cu(II) or which reduce the level of soluble copper, reduce the rate of hydrolysis of PBA. The first order kinetics observed are altered by the presence of diphenylmercury which appears to be a co-catalyst in the hydrolysis of PBA. Hydrolysis of diphenylmercury is achieved only after the elimination of PBA.

Disclaimer

Acknowledgments

The authors wish to recognize and thank Dr. Robert Smiley for insight and valuable discussions of the chemistry of free-radical reactions. We would also like to thank Ms. Cathy Coffey and Ms. Sammie King for their invaluable assistance in performing many long experiments. Additionally we recognize the tireless efforts of the TNX Analytical Laboratory staff and Ms. Shirley King for clerical assistance in preparing this manuscript.

Literature Cited

1. Waters, W. A. *Chemistry of Free Radicals*; 2nd ed.; Oxford: New York, NY, 1948, 162.
2. Migrdichian, V. *Organic Synthesis;* Reinhold: New York, NY, 1957; Vol 2; 1501.
3. Kuivila, H. G.; Nahabedian, K. V. *J. Amer. Chem. Soc.* **1960**, *83*, 2159-63.
4. Kuivila, H. G.; Reuwer Jr., J. F.; Mangravite, J. A. *Can. J. Chem.* **1963**, *41*, 3081-3090.
5. Kuivila, H. G.; Reuwer Jr., J. F.; Mangravite, J. A. *J. Amer. Chem. Soc.* **1964**, *86*, 2666-2670.
6. Cotton, F. A.; Wilkinson, G. *Advanced Inorganic Chemistry;* 5th ed.; John Wiley & Sons: New York, NY, 1988, 327.

7. Smith, R. M.; Martell, A. E. *Critical Stability Constants;* Plenum: New York, NY, 1976, Vol. 4, 47.
8. Ainley, A. D.; Challenger, F. *J. Chem. Soc.* **1930,** 2174-2180.
9. Hu, M. L.; Chang, I. H. *Hu Hseuh Tung Pao* **1965,** 488.
10. Heyrovsky, A. *Anal. Chim. Acta* **1960,** *22,* 405-408.
11. Hodgman, C. D., Ed., *CRC Handbook of Chemistry and Physics;* 36th ed.; Chemical Rubber Publishing: Cleveland, OH, 1954.

RECEIVED January 14, 1994

Chapter 18

Sulfur Polymer Cement as a Final Waste Form for Radioactive Hazardous Wastes

G. Ross Darnell

Idaho National Engineering Laboratory, Transuranic Waste Programs, EG&G Idaho, Inc., Idaho Falls, ID 83415

With ever-increasing emphasis on high-temperature treatment of radioactive and hazardous wastes, the unusual ability of sulfur polymer cement (SPC) to stabilize high loadings of the most troublesome volatilized toxic metals is exciting. SPC is a sulfur polymer composite material that begins melting between 110 and 120°C. Upon cooling it achieves an average compressive strength of 27.6 MPa (4,000 psi), and can triple that strength in 2 years. SPC resists attack by most acids and salts, and its structure suggests long life. We are only on the threshold of developing different formulations of SPC that can accommodate even higher loadings of difficult-to-stabilize wastes. Ongoing SPC tests include stabilizing toxic metals, establishing expected longevity, and evaluating (in nonradioactive full-scale tests) SPC ladened with incinerator ash.

Since there are no known perfect solidification and stabilization agents for radioactive or hazardous wastes, the search continues for individual agents for specific wastes. The U.S. Department of Energy (DOE) began in the early 1980s to test SPC as a radioactive and hazardous waste solidification and stabilization agent because of its unusual properties.

Perhaps SPC's strongest selling point is that it will always melt when elevated to the correct temperature. This feature allows hazardous or radioactive final waste forms that do not pass required tests to be remelted in a low-temperature process and reformulated until the waste forms do pass the tests.

(NOTE: Portland cement has been used for 5 decades as the principal radioactive waste solidification agent, and thus it is accepted as the standard of

0097–6156/94/0554–0299$08.00/0

comparison for other waste solidification and stabilization agents [final waste forms]. Correct titles for the substance addressed in this paper are "modified sulfur cement" and "sulfur polymer cement." For simplicity, the titles are given the single acronym "SPC." The acronym for sulfur polymer cement concrete is "SPCC." The words *cement* and *concrete* refer to the generic family of hydraulic cements and concretes, while *PCC* refers to the specific concrete known as Portland cement concrete. Hazardous waste that is also radioactively contaminated is called mixed waste.)

With waste stabilization and minimization being the primary goals in waste treatment and disposal, DOE's Defense Low-Level Waste Management Program funded tests of SPC ladened with low-level radioactive waste. The research was performed at Brookhaven National Laboratory. DOE's Hazardous Waste Remedial Action Program added funding later. This paper summarizes these and other tests.

Developing and Testing Commercial SPC and SPCC for Harsh Chemical Environs

Pertinent data provided by commercial tests of SPC and SPCC are vital to understanding SPC's potential for stabilizing low-level radioactive, mixed, and hazardous wastes.

Developmental Background. In 1972, the U.S. Bureau of Mines discovered that the addition of dicyclopentadiene and oligomers of cyclopentadiene in equal quantities totaling 5 wt% of the sulfur phase resulted in a construction concrete having advantageous properties not found in other concretes (*1,2,3*). The Bureau of Mines and the Sulphur Institute thereafter joined forces. In 1973, the Sulphur Development Institute of Canada also joined the effort (*4*).

Properties of SPC and SPCC. SPC is a sulfur polymer composite material that begins melting between 110 and 120°C (230 and 248°F), with an optimum pour temperature between 127–138°C (260–280°F). While it takes the average PCC approximately 28 days to achieve a compressive strength of 27.6 MPa (4,000 psi), SPCC reaches that approximate strength upon cooling. In 2 years, its strength can increase by a factor of three (*5*). In remelting SPC, when heat is removed, SPCC will regain its original strength very rapidly as it cools. The various mechanical strengths of SPCC are approximately double those of routine PCC and are not specifically cited herein; however, they are detailed in Reference (*6*). As a proven construction concrete, SPC has demonstrated the ability to survive for years in acids and salts that destroy or severely damage hydraulic concretes in months or even weeks.

In laboratory tests, SPC was found to be impermeable to water (*4*). While both SPC and concrete have approximately the same volume of pores (void space), the pores in SPCC are not connected, whereas the pores in concrete are. The Bureau of Mines worked with the U.S. Environmental Protection Agency (EPA) to protect miners from radon gas. A spraying concept was developed and patented that applied an approximately 7.5-mm (1/4-in.)-thick lining of SPC on mine walls. The SPC lining proved impermeable to radon gas (7).

However, any of the following events will allow both water and gas to penetrate solidified SPCC to varying degrees: (a) if the SPCC is cooled too quickly, it will contain an excess of voids that will connect with each other, (b) if the aggregate in the SPCC contains water, tiny steam vents will develop, or (c) if the wrong aggregate or waste is used, SPCC will become more porous. Fortunately, cooling of the SPCC mass can be controlled with precision (8), water can be routinely eliminated prior to the pour, and the additions of AL_2O_3 or Fe_2O_3 will improve the impermeability of SPCC while reducing the emissions of SO_2 (5). With a given aggregate, SPCC will be less permeable than concrete.

Sulfates attack PCC, but have little or no effect on the integrity of SPCC (1,2). SPCC is corrosion-resistant, and its impermeability protects steel reinforcing materials (and metal waste) from oxidation and subsequent concrete rupture. Where strength and fracture resistance are primary goals, glass fibers, synthetic fibers, epoxy-coated rebar, steel rebar, or a combination thereof can be added (6). Shrinkage, on the average, is 0.1%, slightly greater than PCC. SPCC is resistant to damage by freeze-thaw cycling, and has coefficients of expansion compatible with those of other construction materials, such as concrete and reinforcing steel. Creep in SPC is roughly half that in PCC. Where SPCC and PCC are made with the same aggregate, their densities are nearly identical. Viscosity of SPC is between 25 and 50 cp at 135°C (275°F).

SPCC is best used where exposed to high concentrations of mineral acids, corrosive electrolytes, salt solutions, or corrosive atmospheres in general (5,9). After being exposed to sulfuric acid solutions and copper electrolytic solutions for 9 years, SPCC showed no evidence of corrosion or deterioration. In a 6-year test in a chemical processing plant, PCC was attacked and completely destroyed in some cases, while SPCC showed practically no evidence of strength loss or material degradation. After 7 years of exposure to a salt environment in a test at a potash chemical storage building, two SPCC structural support piers were undamaged, while the PCC pier in the same location was heavily damaged after only 2-1/2 years.

SPCC has its weaknesses in construction applications: it has been shown to deteriorate in hot concentrated chromic acid solutions, hot organic solvent solutions, sodium chlorate-hyperchlorite copper slimes, and strong alkali (over 10%). SPCC is not recommended for placement in areas handling strong bases, strong oxidizing agents, or aromatic or chlorinated solvents (2).

Testing and Modifying SPC for Solidification and Stabilization of Waste

In the 1980s, DOE funded Brookhaven National Laboratory to research, test, and develop SPC to EPA and U.S. Nuclear Regulatory Commission (NRC) requirements. This section discusses those and other efforts.

Tests Completed to EPA Standards. SPC requires no chemical reaction for solidification; it will always solidify when it cools below the melt point and will accept a wide range of waste (aggregate) with divergent chemical and physical compositions. Because of its low viscosity and low-melt temperature, SPC is easier to use than thermoplastics, such as polyethylene (*1*).

Mixed waste fly ash with a pH of 3.8 contained the following (expressed in weight percentages): zinc 36, lead 7.5, sodium 5.5, potassium 2.8, calcium 0.8, copper 0.7, iron 0.5, and cadmium 0.2 (*10,11*). The ash also contained highly soluble metal chloride salts (primarily zinc chloride) that increase the mobility of contaminants while interfering with the solidification reaction of conventional solidification and stabilization agents. The fly ash was combined with SPC and was submitted to the EPA's extraction procedure toxicity (EP Tox) test and to the toxicity characteristic leaching procedure (TCLP). With a loading of 40 wt% fly ash, both cadmium and lead were still above the concentration limits allowed under the Resource Conservation and Recovery Act (RCRA), so additives were sought that would further reduce mobility of the toxic metals. Sodium sulfide (Na_2S) was selected because it reacts preferentially with cadmium and lead (toxic metals) to form highly insoluble metal sulfides. The resultant monolith passed the EPA's EP Tox test and the TCLP, thus meeting the criteria for a delisting petition as hazardous waste.

The need for incineration/vitrification of troublesome wastes before introduction to any solidification agent has long been recognized by the EPA (*12*) and DOE. Toxic metal oxides volatilize in these processes and become secondary waste streams that must be captured and stabilized by another agent. Based on the preceding successes and the need for high-temperature treatment of waste, tests were arranged by the author. Mercury and its compounds are the most troublesome, and volatilization rates can approach 100%. SPC not only serves as an encapsulation material suitable for toxic metal oxides, but also causes chemical conversions from the highly leachable metal oxide form to the unleachable or leach-resistant forms of metal sulfides and sulfates. Heavy loadings (5 wt%) of the eight toxic metals have been combined individually with SPC and 7 wt% sodium sulfide nonahydrate ($Na_2S \cdot 9H_2O$). The leach rates for mercury, lead, chromium, and silver oxides were reduced by six orders of magnitude, while arsenic and barium were reduced by four. Cadmium and selenium were less responsive, with reductions of three and two orders of magnitude, respectively (*13*).

Tests Completed to NRC Standards. The NRC has established a qualifier list of tests under the title "Waste Form Qualification Testing." After the immersion test was completed, compressive strengths of waste-impregnated SPCC ranged from a low of 13.8 MPa (1,998 psi) for 40 wt% boric acid, to a high of 44.4 MPa (6,435 psi) for 40 wt% incinerator ash, with sodium sulfate falling between these extremes. Compressive strength tests after freeze-thaw cycling were well above the NRC limit, and there were some insignificant increases and decreases in strength with different wastes.

The waste contained Co-60 and Cs-137 and was leached for 90 days in compliance with American Nuclear Society, Method 16.1. The leach rate was found to be lower for incinerator hearth ash (Method 14.6) than for highly soluble sulfate salts (Method 9.7). The leach rates were some four to eight orders of magnitude lower than the allowable leach index of 6 established by the NRC. The conclusion was that radionuclides in SPC leach very slowly (*11*).

The NRC tests for biodegradation of SPCC were completed successfully with "no growth" being the result of both the American Society for Testing and Materials (ASTM) G-21 and ASTM G-22 tests. The NRC irradiation test to 10^8 rad was completed successfully. Initially, some deterioration was detected in the SPCC and mixed waste fly ash combination when subjected to the immersion test, but that problem was corrected by the addition of 0.5 wt% glass fibers (*10*).

SPCC specimens containing 80 wt% loadings of lead oxide were successfully subjected to 10^8 rad at Oregon State University as prescribed in the NRC irradiation test. The specimens actually gained 1,000 psi during the irradiation test, and were an order of magnitude better than NRC requirements (McBee, W., Oregon State University, personal communication, 1991). No gas generation was noted. The original intent was to test lead-ladened SPCC as a possible shielding for personnel in high-radiation areas.

In the Netherlands, scientists subjected SPC final waste forms to 10^8 rad and also found that the specimens gained strength. During the irradiation tests, no gaseous radiolysis products were detected (*5*). Inorganic waste and SPC are essentially insensitive to radiation. The constituents that cause radiation effects in the hydraulic or bitumen waste forms are not present in SPC.

The Netherland's Energy Research Foundation ECN, and BNL, dried ion-exchange resins and combined them with SPC. In the worst case, the test specimen crumbled in 3 days by absorbing moisture from the air (60 to 80% humidity). The Netherlands tried another technique. They combined the resins with SPC and raised the temperature 100°C (212°F) to 220–250°C (428-482°F) in the presence of asbestos or diatomite for 3 hours. The resultant test specimen was immersed in water for 1 year and showed no signs of deterioration (*5*).

The author has completed a series of full-scale nonradioactive tests with different mixers in an effort to select one that can be used for the NRC's full-scale tests with mixed waste, and afterwards as an operational SPC mixer (*8*).

Research and development have resulted in a modified SPC that can stabilize waste types like dehydrated boric acid salts, incinerator hearth ash, mixed waste fly ash, and dehydrated sodium sulfate salts that have previously defied solidification and stabilization in concrete in any significant quantity. The modified SPC offers a monolithic waste form that is durable in harsh environments (11).

Comparative test results show that a given volume of waste requires less SPC than cement to achieve a stabilized final waste form that will satisfy EPA and NRC requirements. The following numerical advantages of SPC over cement were calculated: 6.7 times less SPC with sodium sulfate, 3.6 times less SPC with boric acid, 1.4 times less SPC with incinerator bottom ash, and 2.7 times less SPC with incinerator fly ash (11). The average was 3.3 times less SPC than cement.

Since the NRC requires only 500 psi compressive strength, and since the SPCC, which contained various wastes, averaged approximately 4,000 psi (1), there is a large "window of opportunity" for experimentation to develop the optimum SPC mixture for given waste forms.

Waste Streams That Cannot be Combined with SPC for Stabilization. Combining sodium nitrate salts with SPC is not recommended, because, when the two compounds are combined, they could cause a "potentially reactive mixture" (11). SPC is not compatible with highly soluble compounds or organic materials. Expanding clays cannot be used in SPCC for the same reason (2). Chemical corrosion of the final waste form occurs when placed in strong alkaline solutions (above 10%), strong oxidizing solutions like chromic acid and hypochlorite solutions, and some metal slimes like copper.

Safety and Radiation Exposure Considerations

Sulfur is the chemical industry's most widely used raw material (14); therefore, a great deal of information is available on handling it safely. A safe working environment is ensured by following the appropriate procedures provided by the National Safety Council (15), National Fire Protection Association (16), U.S. Department of Health and Human Services (17), National Institute for Occupational Safety and Health, and Manufacturing Chemists' Association, Inc. (18).

Airborne SPC dust can be mildly explosive if all conditions are optimal (normal safety precautions are not exercised); therefore, SPC is procured in pellet or wafer form, which provides handling capabilities with minimal dust creation. SPC and SPCC will emit hydrogen sulfide gas (H_2S) and sulfur vapor to the off-gas system if excessive temperatures are created. Normal heat control systems with backup gas detectors will prevent a safety hazard (15–18). The recommended mixing temperature for SPCC is 127–138°C (260–280°F), which will minimize gaseous emissions to the off-gas system and provide optimum viscosity.

SPC and SPCC are nontoxic (*19*) and do not meet any of the criteria for flammability as established by the United Nations or the U.S. Department of Transportation. SPC's flash point, as determined by the Cleveland open cup method, is 177°C (350°F), and the autoignition is 232–254°C (450–490°F). SPC and SPCC will burn if held in a flame, but they will self-extinguish when the heat is withdrawn (*2,6*). Their poor thermal conductivity (0.2 to 0.5 BTU/h ft °F) is a strong deterrent to melting.

Since SPCC accommodates three to four times more waste than PCC, handling the containers by treatment and disposal operators is reduced, thus reducing the operators' exposure to radiation. SPCC offers self-shielding that also reduces the radiation exposure of operators.

Conclusions

From the beginning, we sought for problems associated with SPC. Thus far, we have been pleasantly surprised at each new promising finding. SPC is a new solidification and stabilization agent (the oldest documented specimens are 15 years old); therefore, continued testing is essential.

All SPC used in tests to date was formulated for the construction industry. The tests discussed, plus many others conducted in countries like Denmark, Japan, France, and Germany, indicate that SPC shows great promise for further development. The search is just beginning for different formulations of SPC that can accommodate higher loadings of various difficult-to-stabilize wastes.

The principal values of SPC are (a) greater waste-to-agent ratio than concrete, (b) the elimination of the need to empty the mixer after a pour, (c) SPC's ability to be remelted and reformulated to pass required tests, (d) less permeability than concrete, (e) the absence of water in the final waste form, (f) SPC's ability to be processed at low temperatures, and (g) less radiation exposure to workers.

With high-temperature processes being the favored treatment for wastes, SPC should be the "clean up" agent that will readily accommodate high concentrations of toxic metals, most notably lead and mercury, in resulting ash and off-gas system residue.

In determining cost and environmental advantages, the volume-reduction factor offered by SPC will be abundantly clear in the ever-increasing high cost of waste disposal. Additionally, the absence of water in the SPCC final waste form offers less chemical breakdown, biodegradation, leaching, and gas generation after disposal. In general, the public stands to benefit significantly both environmentally and economically from using SPC.

Acknowledgments

Work supported by the U.S. Department of Energy, Assistant Secretary for Environmental Restoration and Waste Management, under DOE Idaho Operations Office, Contract DE-AC07-76ID01570.

Literature Cited

1.   Kalb, P. D.; Colombo, P. *Modified Sulfur Cement Solidification of Low-Level Wastes*; BNL 51923, October 1985.
2.   McBee, W. C.; Weber, H. H. "Sulfur Polymer Cement Concrete," *Summary and Proceedings, Twelfth Annual U.S. DOE Low-Level Waste Management Conference, Chicago, Illinois, August 28 and 29, 1990,* CONF-9008119-Proc.
3.   Sullivan, T. A.; McBee, W. C. *Development and Testing of Superior Sulfur Concretes, Report of Investigations 8160*, Bureau of Mines, U.S. Department of the Interior, Washington, D.C., 1976.
4.   McBee, W. C.; Sullivan, T. A.; Fike, H. L. "Sulfur Construction Materials," Bulletin 678, U.S. Department of the Interior, Bureau of Mines, 1985.
5.   Van Dalen, A.; Rijpkema, J. E. *Modified Sulphur Cement: A Low-Porosity Encapsulation Material for Low, Medium, and Alpha Waste,* Commission of European Communities, EUR 12303 EN, 1989.
6.   *National Chempruf Concrete, Inc.,* Chempruf Reference Document, Clarksville, Tennessee, October 1988.
7.   Dale, J. M.; Ludwig, A. C. "Sulfur Coating for Mine Support" (Contract H0210062, Southwest Res. Inst.), Bureau of Mines OFR 31-73, 1972, 46 pp. : NTIS PB 220 408.
8.   Darnell, G. R. *Nonradioactive Full-Scale Tests with Sulphur Polymer Cement (SPC)*, Trip Report, October 8, 1992.
9.   McBee, W. C.; Donahue, D. J. *Corrosion Resistant Concrete*, McBee and Associates, Lebanon, Oregon, undated (approximately 1985).
10.   Kalb, P. D.; Heiser, J. H., III; Colombo, P. *Encapsulation of Mixed Radioactive and Hazardous Waste Contaminated Incinerator Ash in Modified Sulfur Cement*, BNL-43691, undated (approximately 1991).
11.   Kalb, P. D.; Heiser, J. H., III; Colombo, P. "Comparison of Modified Sulfur Cement and Hydraulic Cement for Encapsulation of Radioactive and Mixed Wastes," *Summary and Proceedings, Twelfth Annual U.S. DOE Low-Level Waste Management Conference, Chicago, Illinois, August 28 and 29, 1990,* CONF-9008119-Proc.
12.   U.S. Environmental Protection Agency, *Stabilization/Solidification of CERCLA and RCRA Wastes,* EPA/625/6-89/022, May 1989.
13.   McBee and Associates, "Sulfur Polymer Cement Testing," (Test No. 4 of ongoing tests for EG&G Idaho, Inc.), Subcontract Number C92-170174.
14.   E. R. Johnson Associates, Inc., *An Assessment of the Potential of Modified Sulfur Cement for the Conditioning of Low-Level Radioactive Waste for*

Dumping at Sea, JAI-200, June 1993, prepared for Brookhaven National Laboratory, New York.

15. National Safety Council, "Handling Liquid Sulfur," Chicago, Illinois, Data Sheet I-592, Revision 83, 1983.
16. National Fire Protection Association, Inc., "Standard for Prevention of Sulfur Fires and Explosions," NFPA 655, 1988 edition.
17. U.S. Department of Health and Human Services, Public Health Service Centers of Disease Control, National Institute for Occupational Safety and Health, *NIOSH, Health Hazard Evaluation Report,* HETA 82-292-1358.
18. Manufacturing Chemists' Association, Inc., "Properties and Essential Information for Safe Handling and Use of Sulfur," Chemical Safety Data Sheet SD-74, undated.
19. Tox Monitor Laboratories, Inc., "Plasticized Sulfur 5% DCPD Oligomers," TM 82-356, 1983.

RECEIVED January 14, 1994

Author Index

Affiliation Index

U.S. Department of Agriculture, 223
U.S. Environmental Protection Agency, 51
The University of British Columbia, 114
University of California—Livermore, 8
University of California—Los Alamos, 184
University of Cincinnati, 51

University of Delaware, 197
University of Pittsburgh, 1
The University of Texas at Austin, 96
Wayne State Unversity, 96
Westinghouse Hanford Company, 236
Westinghouse Savannah River Company, 285

Subject Index

A

Abiotic studies, adsorption–desorption
equilibria and kinetics, 54–56
Adsorption–desorption data analysis,
biodegradation kinetics of phenol and
alkylphenols in soils, 71,73–75t
Adsorption–desorption equilibria and kinetics
abiotic studies, 54–56
data analysis for alkylphenols, 71,73–75t
procedure, 61–62
Air streams, gasoline vapor removal by
biofiltration, 143–158
Alkylbenzenes, sorption and
bioavailability studies, 55–56
Alkylphenols in soil, determination of
bioavailability and biodegradation
kinetics, 51–75
Aqueous-based hazardous organic wastes,
treatment by high-energy electron-beam
irradiation, 184
Artificial intelligence approach to synthesis
of waste minimization process
application, 105–112
design philosophy, 97–98
in-plant waste minimization strategy,
98–103
module-based synthesis, 103–106f
Atrazine, oxidation, 223–231

B

Bacteria, role in metal contaminant
mobilization from soil, 78–91
Bacterial–soil sorption isotherm,
determination procedure, 63–64
Bioavailability of phenol and
alkylphenols in soil, 51–75

Biodegradation enhancement of
pentachlorophenol, use of Fenton's
reagent partial oxidation, 197–218
Biodegradation kinetics, information
obtained from quantification, 51
Biodegradation kinetics of phenol and
alkylphenols in soil
abiotic studies on adsorption–desorption
equilibria and kinetics, 54–56
adsorption–desorption data analysis,
71,73–75t
adsorption study procedure, 61–62
bacterial–soil sorption isotherm
determination procedure, 63–64
carbon dioxide evolution rate
measurement procedure, 59–60,65f
CO_2 generation rate
respirometer reactor slurry systems, 67,71
shaker flask, 67,69–70f
$^{14}CO_2$ generation rate, measurement by
radiorespirometry, 71,72f
desorption study procedure, 62
experimental description, 52
future work, 75
kinetic analytical procedure of soil
slurry oxygen uptake data, 57,59
oxygen uptake data analysis, 64,66f–68f,t
radiochemical techniques for
biodegradation determination, 63
respirometric study procedure with soil
slurries, 57
soil microcosm reactor studies,
56–58f,t,64,65f
Biodegradation models for organic
pollutants, development problems, 53
Biofilters, use for waste management, 3
Biofiltration, gasoline vapor removal from
air streams, 143–158

Bestsellers from ACS Books

The ACS Style Guide: A Manual for Authors and Editors
Edited by Janet S. Dodd
264 pp; clothbound ISBN 0–8412–0917–0; paperback ISBN 0–8412–0943–X

The Basics of Technical Communicating
By B. Edward Cain
ACS Professional Reference Book; 198 pp;
clothbound ISBN 0–8412–1451–4; paperback ISBN 0–8412–1452–2

Chemical Activities (student and teacher editions)
By Christie L. Borgford and Lee R. Summerlin
330 pp; spiralbound ISBN 0–8412–1417–4; teacher ed. ISBN 0–8412–1416–6

Chemical Demonstrations: A Sourcebook for Teachers,
Volumes 1 and 2, Second Edition
Volume 1 by Lee R. Summerlin and James L. Ealy, Jr.;
Vol. 1, 198 pp; spiralbound ISBN 0–8412–1481–6;
Volume 2 by Lee R. Summerlin, Christie L. Borgford, and Julie B. Ealy
Vol. 2, 234 pp; spiralbound ISBN 0–8412–1535–9

Chemistry and Crime: From Sherlock Holmes to Today's Courtroom
Edited by Samuel M. Gerber
135 pp; clothbound ISBN 0–8412–0784–4; paperback ISBN 0–8412–0785–2

Writing the Laboratory Notebook
By Howard M. Kanare
145 pp; clothbound ISBN 0–8412–0906–5; paperback ISBN 0–8412–0933–2

Developing a Chemical Hygiene Plan
By Jay A. Young, Warren K. Kingsley, and George H. Wahl, Jr.
paperback ISBN 0–8412–1876–5

Introduction to Microwave Sample Preparation: Theory and Practice
Edited by H. M. Kingston and Lois B. Jassie
263 pp; clothbound ISBN 0–8412–1450–6

Principles of Environmental Sampling
Edited by Lawrence H. Keith
ACS Professional Reference Book; 458 pp;
clothbound ISBN 0–8412–1173–6; paperback ISBN 0–8412–1437–9

Biotechnology and Materials Science: Chemistry for the Future
Edited by Mary L. Good (Jacqueline K. Barton, Associate Editor)
135 pp; clothbound ISBN 0–8412–1472–7; paperback ISBN 0–8412–1473–5

For further information and a free catalog of ACS books, contact:
American Chemical Society
Distribution Office, Department 225
1155 16th Street, NW, Washington, DC 20036
Telephone 800–227–5558